Evolutionary ecology of plant-plant interactions

Grethe Trillingsgård
12.06.36 – 1.12.93

Evolutionary ecology of plant-plant interactions

An empirical modelling approach

Christian Damgaard

AARHUS UNIVERSITY PRESS

Evolutionary ecology of plant-plant interactions
An empirical modelling approach

© Copyright: Christian Damgaard and Aarhus University Press 2004
Cover design: Tinna Christensen and Kathe Møgelvang, National Environmental Research Institute
Layout: Tinna Christensen and Kathe Møgelvang, National Environmental Research Institute
Paper: 100 g Munken Pure
Printed by Narayana Press, Gylling

ISBN 87-7934-116-0

Christian Damgaard
Department of Terrestrial Ecology
National Environmental Research Institute
Vejlsøvej 25
DK-8600 Silkeborg
www.dmu.dk

AARHUS UNIVERSITY PRESS
Langelandsgade 177
DK-8200 Aarhus N
Fax +45 89 42 53 80
www.unipress.dk

Published with financial support by
Danish Natural Science Research Council

Preface

Evolutionary ecology

Evolutionary biology and ecology share the goals of describing variation in natural systems and understanding its functional basis. Within this common framework, evolutionary biologists principally describe the historical lineage-dependent processes, while ecologist focus on the contemporary processes. This difference is summarised in the commonly used truism that evolutionary time scales are longer than ecological time scales. Evolutionary ecologists try to integrate the two approaches by studying variation at all levels, from variation between individuals to variation among communities or major taxonomic groups.

Traditionally the mathematical formulation of evolutionary and ecological problems has differed in one important aspect, which is the description of individuals. In evolutionary or population genetic models an individual plant is expressed relatively to the total size of the population as a frequency, whereas individuals in ecological models are counted in absolute numbers or per area as a density. Mathematically it is easier to work with frequencies. However, for many ecological problems mathematical models expressed in frequencies are too degenerated to provide suitable solutions. For example, it is a well known fact that in a world with limited resources all population growth has to stop at a certain point, however, this fact cannot be expressed in a traditional population genetic model, which implicitly assumes permanent exponential growth.

Integrating ecological data and mathematical modelling

Evolutionary biology has since long been a quantitative scientific discipline. However, there has been a strong tradition in plant ecology to describe different plant communities and succession processes in a qualitative way. Possibly due to the obvious importance of plasticity and the spatial setting, which only lately has started to be incorporated in the ecological models, many field ecologists have felt that mathematical modelling have had little to offer in their attempt to understand the dynamics of plant communities. As a consequence of the lack of communication between field ecologists and mathematical modellers, many ecological studies have been inappropriately analysed with standard linear models and, on the other hand many mathematical modellers have tended to examine parameter space rather than ecological data.

However, there has been a growing interest to make simple and at the same time more biologically realistic plant ecological models, and due to the powerful computers it is now possible to fit ecological data to such simple ecological models with biological meaningful parameters. This will allow a more rigorous testing of various ecological hypotheses and the development of quantitative ecological predictions. Such predictions are highly demanded both by the public, e.g., in conservation management and risk assessment of genetically modified plants, and in order to advance the scientific field of plant ecology (Keddy 1990, Cousens 2001).

It seems that the dialog between ecologists and mathematical modellers, which have proved so fruitful in other areas of ecology, is now also beginning to develop in plant ecology. It is my hope that this monograph will further strengthen the bond between plant ecology and modelling.

Outline of monograph

This monograph will discuss and develop concepts and simple empirical models that are useful in the study of quantitative aspects of the evolutionary ecology of plant – plant interactions and the statistical analysis of plant ecological data. Special attention will be paid to the consequences of the sedentary life form of adult plants and the subsequent strong interactions between neighbouring plants. The monograph will provide an overview of different evolutionary and ecological empirical plant population models and provide conceptual links between different modelling approaches, e.g., spatial individual-based or plant size explicit modelling and the equilibrium conditions of mean-field models. The biological information underlying the discussed models will be summarised. However, it is not the scope to present a full discussion of the biology of plant – plant interactions, which have been treated extensively by other authors (e.g., Harper 1977, Grime 2001, Silvertown and Charlesworth 2001).

The consequences of a sedentary life history with strong interactions with neighbouring plants will be introduced in chapter 1. Single-species competitive plant growth models will be described in chapter 2, where the growth of individual plants is modelled with increasing complexity as functions of plant size, plant density, and the spatial distributions of plants. In chapter 3, models describing the demography of a single plant species, including mortality, reproduction, seed dispersal and dormancy will be discussed and linked to different population growth models and equilibrium conditions. After ecological concepts and models are introduced in the single-species case, modelling of the interactions between species will be introduced in chapter 4, where emphasis will be on equilibrium conditions and how to predict the probability of different ecological scenarios as a function of the environment. The on-going discussion on the ecological success of different plant strategies will be introduced and put into a modelling context. In chapter 5, the genetic analysis of population structure will be introduced and the effect of inbreeding and finite population sizes on the genetic variation will be discussed. In chapter 6, one-locus sex asymmetric and density-dependent mixed-mating selection models, which are particular relevant for plant populations, will briefly be introduced, after which the measuring of natural selection will be discussed. In chapter 7, different genetically and ecologically based hypotheses on the evolution of plant life history will briefly be discussed. Finally, in the appendices there is a list of the parameters with a fixed usage in chapters 2-4 (A), an introduction to linear regression technique (B), Bayesian statistics (C), and the stability analysis of discrete dynamic systems (D). Mathematica notebooks exemplifying the methodology introduced in this monograph may be downloaded from my webpage.

This monograph was written as a part of my doctoral thesis at Aarhus University, where I present my scientific contributions over the last decade in a coherent way. Consequently the monograph is a reflection of my views on the issues rather than a balanced account of the field and the cited references are somewhat biased towards my own production.

Acknowledgements
Thanks to Liselotte Wesley Andersen, Malene Brodersen, Lene Birksø Bødskov, Tinna Christensen, Marianne Erneberg, Gösta Kjellsson, Christian Kjær, Hans Løkke, Kathe Møgelvang, Birgit Nielsen, Vibeke Simonsen, and Morten Strandberg for critical comments and help and Jacob Weiner for indispensable collaboration. The Carlsberg Foundation supported the writing of this monograph.

Content

1. Introduction

Ecological and evolutionary success

How may we characterise, measure and predict the ecological and evolutionary success of a plant species? Intuitively, ecological success may be characterised as the presence rather than absence of a plant species under specific conditions, and it may be simply measured by the probability that a plant species is present under these specific conditions. Likewise, evolutionary success may be characterised by the presence rather than absence of offspring from a common ancestor. The evolutionary success of a genotype is measured by its fitness, i.e. the number of successful offspring.

In order to predict the ecological success of a plant species or fitness of a genotype we need to describe the complicated biological processes taking place in a plant life. Generally speaking a terrestrial plant is comprised of roots that provides access to water and nutrients, vegetative parts that provides a structure for converting solar energy into biomass, defence structures to minimise the potential damage off herbivores and pathogens, and reproductive structures that provides an opportunity for reproduction and dispersal of offspring. The nature of the different structures and the allocation of resources among the different struc-

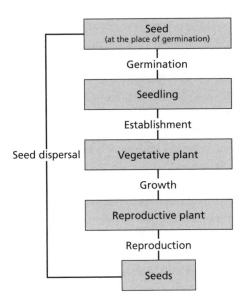

Fig. 1.1 Life stages (in boxes) and processes for a typical annual plant.

tures determine the ecological success. Additionally, the life history of plants is comprised by a number of stages: germination, establishment, growth, reproduction, and dispersal (Fig. 1.1), and the characteristics of the plant during these stages also has a large influence on the ecological success.

None of the above-mentioned plant characteristics are fixed properties of a plant species. Some of the characters show allometric responses, i.e., they depend on the size of the plant. Most plant characters have some degree of plasticity on the individual plant level as a response to the abiotic and biotic environment. For example, the allocation between roots and vegetative parts and the probability of germination may change as a response to the availability of water. The plant characteristics may also change over time at the population level as a response to natural selection in a changing environment.

The consequences of being sedentary

A notable character of the majority of terrestrial plants is that the adults are primarily sedentary. Some plant species may send vegetative runners out in different directions, but most species stay more or less at the place of germination. This sedentary life form has a profound impact on the life history of individual plants, as well as influencing the ecology and evolution of plant populations. Plants that are stuck at the place of germination experience competition from larger neighbouring plants more severely than animals that may move to an environment with less competition. It is essential that the description of the ecological and evolutionary forces of plant populations and communities incorporate the consequences of the sedentary adult life form and not just adopt the available descriptions of free-moving animal populations uncritically.

Neighbouring plants affect all the different processes in the life history of plants (Fig. 1.1). The quantity and quality of light, which is modified by neighbouring plants, or chemical substances excreted from neighbouring plants (allelopathy) may influence seed germination. The establishment of seedlings from a germinated seed depends for many plant species critically on some sort of disturbance of the soil that relieves the competition from neighbouring plants.

Neighbouring plants are one of the most important universally biotic limiting factors of plant establishment and growth. Negative plant-plant interactions seems to be important in most environments in controlling the growth of individual plants, although, positive plant-plant interactions have been reported in extreme environments (Callaway 1995, Stoll and Weiner 2000). Other biotic factors that controls plant establishment and growth include herbivores and pathogens, but their role is more variable than the negative competitive effect from neighbouring plants (Crawley and Pacala 1991).

As a possible adaptation to the sedentary life form and severe competition for the limiting resources (light, water or nutrients) from neighbours, the same plant genotype may show a large variation in phenotypic characters. Especially the number of different plant parts, i.e. branches, leaves, flowers, fruits etc., may vary considerably (Harper 1977), and in plant populations at medium or high densities where will often be a large size variation among conspecific individuals of approximately the same age (Weiner 1990).

The fecundity (seed production) is usually positively correlated with the size of the plant, which again is controlled by the density of neighbouring plants and the abiotic environment. The male reproductive fitness in wind - or insect pollinated outcrossing plants often depends on the density of conspecific neighbouring plants (Lloyd and Bawa 1984). A partial self-fertilising mating system, which influences the genetic structure and the evolutionary processes, is common in plant populations (Schemske and Lande 1985). Self-fertilisation may also be looked upon as an adaptation to the sedentary life form; after a colonisation event, isolated sexual plants may have difficulties to receive compatible pollen if they are self-incompatible. With self-compatible individuals a single propagule is sufficient to start a sexually reproducing population, making its establishment much more likely than if the chance growth of two self-incompatible individuals sufficiently close together spatially and temporally is required (Baker 1955).

A necessary adaptation to the sedentary adult life form is seed and/or vegetative dispersal. Most plant communities are dynamic with continuous local disturbances followed by a relatively long succession process where plant species have to be able to re-colonise a local area. On a larger time scale, plant species have to colonise new areas due to the recurring environmental changes, e.g., temperature shifts during ice ages. Successional dynamics after a disturbance event are to a certain degree predictable. Early-successional plants species typically have a series of correlated traits, including high fecundity, long-distance dispersal, rapid growth when resources are abundant and slow growth and low survivorship when resources are scarce. Late-successional species usually have the opposite traits, including relatively low fecundity, short dispersal distances, slow growth, and an ability to grow, survive, and compete under resource-poor conditions (Grime 2001, Rees et al. 2001). It has been suggested that much of the plant species diversity among plant communities is controlled by a trade-off between the ability to colonise new habitats and the ability to compete for resources (Rees et al. 2001). Furthermore, the evolution of seed size and consequently seed dispersal is among other factors affected by neighbouring plants (Geritz et al. 1999), and the species composition of the neighbouring plants may influence seed dispersal by affecting wind speed or influencing the number and species composition and behaviour of possible animal vectors (Harper 1977).

Modelling plant-plant interactions

It is possible to test simple ecological and evolutionary hypotheses using informal and verbal models, but verbal models may be susceptible to vagueness and logical pitfalls, especially if the model is complicated. On the other hand, by constructing a mathematical model, which encapsulates the hypothesis that we want to test, we are forced to be precise in the description of the functional relationships.

Biological systems are variable and involve some degree of uncertainty; and when the biological processes are modelled mathematically it is important to take this uncertainty into account. Two fundamentally different approaches can be taken when modelling plant-plant interactions and the associated stochastic variation:

1. Using *mechanistic* plant models, the expected effects of the abiotic and biotic environment, including effects of neighbouring plants, on plant growth and reproduction may be modelled. In a mechanistic plant model (e.g., Tilman 1988, Pacala et al. 1996, Deutschman et al. 1997, Cernusca et al. 1999), the growth rate of different plant species is modelled as a function of the limiting resource(s). Often, but not always, the models are stochastic models that describe the plant growth of individual plants in a specific spatial setting. Such mechanistic plant models ideally contain detailed information of how plants compete for resources and respond to different environments, which enable predictions of plant growth as a function of the environment. However, the mechanistic models necessarily contain many parameters that are difficult or impossible to estimate jointly and the use of these models for ecological predictions is questionably (Levin et al. 1997). Thus even though ecological predictions and tests of ecological hypotheses often are the alleged objectives of these mechanistic models their main importance is probably to test *our understanding* of plant population and community-level processes. In the words of Levin et al. (1997): "...these models produce cartoons

that may look like nature but represent no real systems. However they do represent powerful experimental tools, which become more valuable when used to produce exhaustive simulations that allow exploration of parameter space and model structures".

2. Using *empirical* (or phenomenological) plant models that are fitted to plant growth data in a statistical sense (Appendix B). Empirical models are mathematically simple deterministic models with relatively few biologically interpretable parameters and due to their simplicity it is possible to jointly estimate all the parameters in the model from a single data set, which allow estimation of any covariability between parameters. The empirical models rely heavily on data and the model will never be better than the underlying data, thus it is important to be conscious of the limitations of empirical models, e.g., it is not possible to make predictions outside the domain of the data. Since ecological data typically has been obtained at a fixed environment, empirical models have been criticised for having limited predictive power, however obtaining ecological data along an environmental gradient may circumvent this limitation. In constructing the empirical ecological models, there is a delicate balance between keeping models simple and describing the biological system in sufficient detail to obtain the desired information, i.e., performing tests of investigated hypotheses or obtaining posterior distributions of biological parameters (Appendix C).

In this monograph the empirical modelling approach will be followed almost throughout. However, the two modelling approaches may be linked by making an extensive sensitivity analysis of the parameters and the dynamics in a mechanistic plant model, and try to reduce the mechanistic plant model to a mathematically simpler model while maintaining the important dynamics and parameters (Levin et al. 1997).

2. Individual plant growth

Competitive growth

The energy requirement of a plant for maintenance and growth depends critically on photosynthesis that again depends on available water, carbon dioxide, and sunlight of a sufficient quality, i.e. short-wave solar radiation. Simplistically it may be said that, factors such as temperature, water and nutrient availability in many environments determine the potential for growth, but growth is only realised if the individual plant receives sufficient light.

In many environments neighbouring plants limit the growth of each other because they compete for the access to a specific limiting resource (Fig. 2.1). The mode of competition depends on the resource and the size of the competing plants. Some resources like light of short wavelength, which is depleted approximately exponentially by successive leaf layers, may be monopolised by tall plants (monopolising competition). Other resources, like phosphor, are more or less evenly distributed in the soil, so that a large root system cannot prevent a small root system of phosphor uptake (exploitative competition). For resources that may be monopolised larger plants may be able to obtain more of the resource than their share, based on relative size, and to suppress the growth of smaller individuals (Begon 1984). This positive effect of size on the competitive ability of a plant is known as size-asymmetric competition, although often used synonyms in the ecological literature include, "one-sided competition", "contest competition", or "dominance-suppression competition".

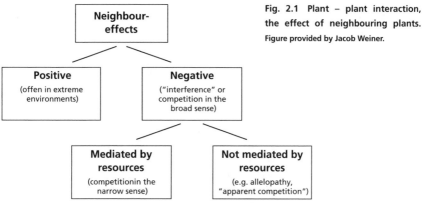

Fig. 2.1 Plant – plant interaction, the effect of neighbouring plants. Figure provided by Jacob Weiner.

In order to understand the consequences of resource competition on plant growth, ideally we would like to know the flow of the resources between the many compartments in the complicated three-dimensional space where plants grow. However, at this point in time competition for resources among terrestrial plants has to be inferred indirectly by studying the growth of plants at different environments, densities and spatial arrangements.

In an elegant experiment with morning-glory vines (*Ipomoea tricolor*) competition for light among the shoots was separated from competition for water and nutrients among the roots (Weiner 1986) (Fig. 2.2). The effect of root competition was most severe when measured by the decrease in *mean* plant weight and comparable to the decrease in mean plant weight, when both shoots and roots were competing indicating that either water or nutrients was the limiting resource. However, when the vines competed for light there was a large *variation* in plant weight indicating size-asymmetric competition for light. Furthermore, the coefficient of variation in plant weight when shoots were competing was comparable to the coefficient of variation when shoots and roots were competing, suggesting that the size-asymmetry of competitive interactions may be determined by a resource that is not the limiting resource of the population. In the case of morning-glory vines water or nutrients were the limiting resources for the growth of the vine *population*, whereas light was the limiting resource for some of the *individual* vine plants.

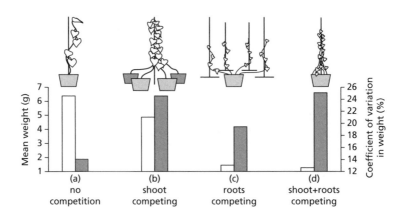

Fig. 2.2   An experiment where root and shoot competition were separated using vines (*Ipomoea tricolor*) (Weiner 1986). Gini Coefficients (see later): a: 0.081, b: 0.139, c: 0.112, d: 0.143. Mean shoot dry weight (open bars) was significantly different for all comparisons between treatments except (c) and (d). The CV in shoot dry weight (shaded bars) and the Gini Coefficients were significantly different for the comparison of treatments (a) and (b) and for (a) and (d). Figure after Weiner (1990).

Generally, competition does not seem to be size-asymmetric when plants have a short growth period, grow on very poor soils, at low density, or competing only for below ground resources (Schwinning and Weiner 1998). But the degree of size-asymmetric competition varies over a continuum from complete symmetry, where all plants irrespective of their size receive the same amount of resource, to complete size-asymmetry, where the large plant get all the contested resource (Schwinning and Weiner 1998).

Describing variation in plant size

The first step in understanding the size variation obtained in plant competition experiments is to describe the variation that is studied. Variation in plant size has traditionally been described and analysed using the statistical moments (mean, variance etc.) of the size distribution, or statistics derived from the moments such as standard deviation and skewness. In recent years, the focus has shifted towards an emphasis on inequality in size in being a more ecological relevant measure, after it was argued that "size variability" or "size hierarchy", as the terms are used by ecologists, are often synonymous with the concept of size inequality. One approach to inequality that has been applied to plant populations is the Lorenz curve (Weiner and Solbrig 1984).

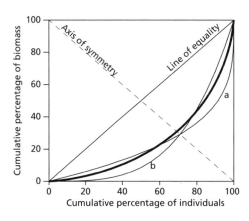

Fig. 2.3 Three Lorenz curves: a symmetric case (bold line), and two size-asymmetric cases (a, b).

In the Lorenz curve, individuals are ranked by size, and the cumulative proportion of plants (x-axis in Fig. 2.3) is plotted against the corresponding cumulative proportion of their total size (y-axis in Fig. 2.3). If we have a sample of n ordered plants, so that v'_i is the size of plant i, and $v'_1 \leq v'_2 \leq \cdots v'_n$, then the sample Lorenz curve is the polygon joining the points $(h/n, L_h/L_n)$, where $h = 0,1, \cdots$, and $L_0 = 0$, and $L_h = \sum_{i=1}^{h} v'_i$ (Kotz et al. 1983).

Alternatively, the Lorenz curve can be expressed as

$$L(y) = \int_0^y v \, dF(v)/\mu \tag{2.1},$$

where $F(y)$ is the cumulative distribution function of ordered plants, and μ is the average plant size (Dagum 1980, Kotz et al. 1983).

If all individuals are the same size, the Lorenz curve is a straight diagonal line, called the line of equality (Fig. 2.3). If there is any inequality in size, then the Lorenz curve is below the line of equality. The total amount of inequality can be summarised by the Gini Coefficient (or Gini Ratio), which is the ratio between the area enclosed by the line of equality and the Lorenz curve, and the total triangular area under the line of equality (Fig. 2.3). The Gini Coefficient is most easily calculated from unordered plant size data as the "relative mean difference", i.e., mean of the difference between every possible pair of individuals, divided by the mean size (Sen 1973):

$$G = \frac{n}{n-1} \cdot \frac{\sum_{i=1}^{n} \sum_{j=1}^{n} |v_i - v_j|}{2n^2\mu} \tag{2.2},$$

or alternatively, if the data is ordered by increasing plant size (Dixon et al. 1987):

$$G = \frac{n}{n-1} \cdot \frac{\sum_{i=1}^{n} (2i - n - 1)v'_i}{n^2\mu} \tag{2.3},$$

where $n/(n-1)$ is a bias correction (Glasser 1962). The Gini Coefficient ranges from a minimum value of zero, when all individuals are equal, to a theoretical maximum of one in an infinite population in which every individual except one has a size of zero. The Gini Coefficient has been used as a measure of inequality in size and fecundity in plant populations in numerous studies (e.g. Weiner 1985, Geber 1989, Knox et al. 1989, Preston 1998).

Like any summary statistic, the Gini Coefficient does not contain all the information in the Lorenz curve, and it has been pointed out that different Lorenz curves can have the same Gini Coefficient (Weiner and Solbrig 1984, Shumway and Koide 1995). In case a of Fig. 2.3, most of the inequality within the population is due to the few largest individuals, which contain a large percentage of the population's biomass. In case b, the same overall degree of inequality is due primarily to the relatively large number of small individuals, which are contributing little to the population's total biomass. This difference can be quantified by measuring the asymmetry of the Lorenz curve around the other diagonal (axis of symmetry in Fig. 2.3), specifically the location of the point at which the Lorenz curve has a slope equal to 1.

The asymmetry of the Lorenz curve may be quantified using the Lorenz Asymmetry Coefficient, which is defined as $S = F(\hat{\mu}) + L(\hat{\mu})$, where the functions F and L are as in equation (2.1) (Damgaard and Weiner 2000). A Lorenz curve is symmetric if the curve is parallel with the line of equality at the axis of symmetry, and since the axis of symmetry can be expressed by $F(y) + L(y) = 1$, we have that a Lorenz curve is symmetric if and only if $S = F(\hat{\mu}) + L(\hat{\mu}) = 1$ (Kotz et al. 1983, Damgaard and Weiner 2000). If $S > 1$, then the point where the Lorenz curve is parallel with the line of equality is above the axis of symmetry. Correspondingly, if $S < 1$, then the point where the Lorenz curve is parallel with the line of equality is below the axis of symmetry.

Since the sample Lorenz curve is a polygon, we can calculate the sample statistic S from the ordered plant size data using the following equations (Damgaard and Weiner 2000):

$$\delta = \frac{\hat{\mu} - v'_m}{v'_{m+1} - v'_m}$$ (2.4),

$$F(\hat{\mu}) = \frac{m + \delta}{n}$$ (2.5),

$$L(\hat{\mu}) = \frac{L_m + \delta v'_{m+1}}{L_n}$$ (2.6),

where δ is a correction factor for finite samples and m is the number of plants with a plant size less than $\hat{\mu}$. Confidence intervals for the estimates of S for a sample can be obtained with the bootstrapping procedure, as has previously been demonstrated for the Gini Coefficient (Dixon et al. 1987).

Modelling plant growth

Above it was explained how the size variation obtained in plant competition experiment may be described. The next step in the understanding of the effect of neighbouring plants is to compare and test the different hypotheses of the effect of neighbours that may have arisen during the description of the experimental data.

Several sigmoidal (or saturated) empirical growth models have been proposed to describe growth of individual plants, e.g., Gompertz, and logistic growth models where the main difference between the models is when the plant experiences its maximum growth rate (Seber and Wild 1989). The maximum growth rate occurs at the point when the sigmoidal growth curve shifts from being convex to being concave (the inflection point). In the Richards growth model (Richards 1959, Vandermeer 1989, Garcia-Barrios et al. 2001) this inflection point is modelled by a free parameter. The determination of the inflection point by a free parameter makes the Richards growth model relatively flexible and inclusive of the other sigmoidal growth models (Fig. 2.4). Rather than assuming a fixed point of inflection, the Richards growth model is assumed in the following, since there is no general theory that predicts at what growth state plants experience their maximum growth rate, and the inflection point has been shown to depend on density (Damgaard et al. 2002).

Individual plant growth will in the following be measured by the absolute growth rate, which is the increase in some measure of plant size, e.g., biomass, per time. In the Richards growth model, the growth of a plant at time t is assumed to be proportional to the plant size at time t multiplied by a saturating function of the plant size at time t.

$$\frac{dv(t)}{dt} = \frac{\kappa}{1-\delta}\, v(t)\left(\left(\frac{v(t)}{w}\right)^{\delta-1} - 1\right) \quad \delta \neq 1 \tag{2.7},$$

where $v(t)$ is the plant size at time t, κ is a growth parameter, and w is the final plant size. The initial growth rate is $\kappa/(\delta-1)$ and if $\delta > 1$ the initial growth is exponential. The shape of the growth curve is mainly determined by δ. If $\delta > 0$, then the growth curve is sigmoidal and the point of inflection is at the proportion $\delta^{1/(1-\delta)}$ of the final size, i.e., the size at which plant growth starts to deviate from exponential growth decreases with δ (Fig. 2.4). The slope of the tangent at the point of inflection decreases also with δ. The plant experiences the maximum growth rate, $\kappa w \delta^{\delta/(1-\delta)}$, at the inflection point. The saturating term is analogous with the competition term in the logistic model of population growth of individuals towards a carrying capacity (Verhulst 1838).

The variable inflection point of the Richards growth model includes other sigmoidal growth functions as special cases (Fig. 2.4) (Seber and Wild 1989), i.e., the monomolecular model ($\delta = 0$), the von Bertalanffy model ($\delta = 2/3$), the logistic model ($\delta = 2$), and by taking the limit as $\delta \to 1$ the Gompertz model:

$$\frac{dv(t)}{dt} = \kappa v(t)(\log(w) - \log(v(t))) \tag{2.8}.$$

The Richards growth differential equation may be solved under the condition that for $\delta < 1$, $(1 - \delta)\exp(\kappa \gamma) \leq 1$ (Seber and Wild 1989):

$$v(t) = w(1 + (\delta - 1)\exp(-\kappa(t - \gamma)))^{1/(1-\delta)} \tag{2.9},$$

where γ is the time of the inflection point.

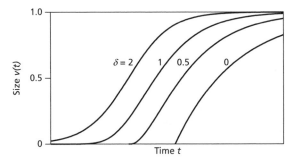

Fig 2.4 Curves of Richards growth model for various δ. The final plant size, $w = 1$ and κ is 0.75, 0.5, 0.375 and 0.25, respectively. The parameter γ is changed each time so that the curves do not sit on top of one another. The variable inflection point controlled by the parameter δ makes the Richards growth model inclusive of the other sigmoidal growth functions, e.g., the logistic model ($\delta = 2$), the Gompertz model ($\delta = 1$), and the monomolecular model ($\delta = 0$). Figure after Richards (1959).

Size-asymmetric growth

The Richards growth model (2.7) adequately describes the growth of a single plant or plant growth in a monoculture of identical plants. However, the plants in a monoculture are rarely identical. There may be variation in the time of germination, distance to nearest neighbour, or in the microenvironment that leads to variable plant sizes. If the plant growth is limited by a resource that may be monopolised, e.g., light, then size-asymmetric competition may occur and the variation among plant sizes may increase with time more than expected under Richards growth model.

The effect of size-asymmetric growth may be included in the Richards growth model by modelling individual plant growth as proportional to a power function of their size (Schwinning and Fox 1995, Damgaard 1999, Wyszomirski et al. 1999, Damgaard et al. 2002)

$$f(v(t), a) = \begin{cases} 1 & a = 0 \\ (v(t) + 1)^a - 1 & a > 0 \\ 1 \text{ or } 0 & a = \infty \end{cases} \tag{2.10}$$

where the effect of plant size on growth is summarised by a size-asymmetry parameter, a, which measures the degree of curvature of the size-growth relationship over the entire growth curve and takes values between 0 and ∞ (Table 2.1, Fig. 2.5). The reason for choosing a power function of plant size in (2.10) is somewhat arbitrary but may be motivated by the favourable scaling properties of the power function.

In order to take the effect of plant size variation on the growth of individual plants into account, an individual-based Richards growth model may be formulated by generalising (2.7) and (2.8) with respect to size-asymmetric growth (2.10). Assume a monoculture of n competitively interacting plants of variable size, then the growth of plant i at time t may be expressed by n coupled differential equations,

Table 2.1   Classification of the degree of size-asymmetry based on the relationship between size and growth rate. Terminology adapted after (Schwinning and Weiner 1998).

Parameter value		Definition
Complete symmetry	$a = 0$	All plants have the same growth rate irrespective of their size.
Partial size-symmetry	$0 < a < 1$	The growth rate is less than proportional to the size of the plant.
Perfect size-symmetry	$a = 1$	The growth rate is proportional to the size of the plant.
Partial size-asymmetry	$a > 1$	The growth rate is more than proportional to the size.
Complete size-asymmetry	$a = \infty$	Limiting case where only a few dominating plants are growing; all other plants have stopped growing.

$$\frac{dv_i(t)}{dt} \left\{ \begin{array}{l} \dfrac{\kappa}{1-\delta}\, f(v_i(t),\, a)\left(\left(\dfrac{1}{n\,w}\sum_{j=1}^{n} v_j(t)\right)^{\delta-1} -1\right) \quad \delta \neq 1 \\[2em] \kappa f(v_i(t),\, a)\left(\log(nw) - \log\left(\sum_{j=1}^{n} v_j(t)\right)\right) \quad \delta = 1 \end{array} \right. \qquad (2.11),$$

(Damgaard et al. 2002), where $v_i(t)$ is the size of plant i at time t, and w is the *average* plant size at the end of the growing season. When $\delta > 1$, the initial growth rate is $(\kappa a)/(\delta - 1)$.

The saturating term $\left(\dfrac{1}{n\,w}\displaystyle\sum_{j=1}^{n} v_j(t)\right)^{\delta-1} -1$ or $\left(\log(n\,w) - \log\left(\displaystyle\sum_{j=1}^{n} v_j(t)\right)\right)$

if $\delta = 1$, measures the decrease in individual plant growth due to the size and competitive effects of the n interacting plants. The saturation term, which is equal for all n plants at a given time, reduces as the plants grow and when $\sum_{j=1}^{n} v_j(t) = nw$ the saturation term equals zero and growth stops.

The individual-based Richards growth model cannot be solved in the general case, and in order to fit the growth model to growth data the model has to be solved numerically for each set of parameter values used in a maximum likelihood fitting procedure (Damgaard et al. 2002).

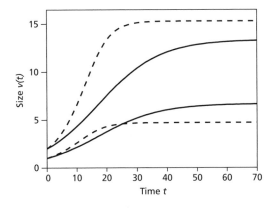

Fig 2.5 Individual-based Richards growth model for two interacting plants with an initial size of one and two, respectively, at two levels of the degree of size-asymmetry.
Initial growth rate: $(\kappa a)/(\delta - 1) = 0.1$.
Full line: $a = 1$, $w = 10$, $\kappa = 0.1$, $\delta = 2$.
Dashed line: $a = 1.5$, $w = 10$, $\kappa = 0.0667$, $\delta = 2$. Notice that the size difference increases with a.

Example 2.1 Size-asymmetric growth in
Chenopodium album

Chenopodium album (Chenopodiaceae) is a broad-leaved summer annual that colonises disturbed habitats and is a common weed in the
farmland. The individual-based Richards growth model (2.11) was fitted to growth data from an experiment with *C. album* monocultures
grown in an experimental garden at different densities, where the
height and diameter of individual plants were measured non-destruc-
tively nine times during the growth season (Table 2.2) (Nagashima et al.
1995, Damgaard et al. 2002). The estimated average size of the plants
at the end of the growing season (*w*) decreased significantly at higher
density. The estimate of the size-asymmetry parameter (*a*) was always
significantly greater than one and it increased significantly at higher
density. Thus, competition in these crowded monocultures of *C. album*
was found to be size-asymmetric and the degree of size-asymmetry
increased at higher density. The growth of *C. album* individuals fitted
the individual-based Richards growth model (2.11) significantly better
than simpler growth models with one parameter less (e.g., generalised
logistic ($\delta = 2$) or Gompertz ($\delta = 1$) growth models) and density had a
significant effect on this shape parameter, i.e., increasing density decreased the size at which plants began to deviate from exponential
growth. It may be concluded that the additional parameter in the individual-based Richards growth model, which determines the location of
the inflection point of the growth curve, provides necessary flexibility
in fitting growth curves.

Table 2.2 Maximum likelihood estimates and 95% credibility intervals (see
Appendix C) of the parameters in the individual-based Richards growth model
(2.11) fitted to *Chenopodium album* growth data (Nagashima et al. 1995). The
biomass of plants was assumed to be proportional to height × stem diameter^2.
Plants that died no longer contribute to the population's biomass (i.e. they no
longer compete with living plants). The procedures of fitting the growth data to
the model are explained in detail in Damgaard et al. (2002).

Plant density	κ	a	w	δ
400	0.0160	1.16	2460	0.80
	(0.0104 – 0.0324)	(1.07 - 1.22)	(2200 – 2780)	(0.72 – 0.95)
800	0.0046	1.31	1070	0.43
	(0.0023 – 0.0062)	(1.27 – 1.39)	(1030 - 1110)	(0.33 – 0.55)

Effect of plant density

In Chenopodium album it was found that the average final plant size decreased with density and the degree of size-asymmetric competition increased with density (Table 2.2), and generally it has been found that the effect of plant-plant interactions increase with plant density (Harper 1977). The negative relationship between plant size and plant density due to competitive interactions is probably one of the best-studied aspects in plant ecology.

In a number of empirical studies (e.g., Shinozaki and Kira 1956, Law and Watkinson 1987, Cousens 1991) the class of hyperbolic size-density response functions has been demonstrated to fit plant competition data well at different growth stages and for different measures of plant size. Bleasdale and Nelder (1960) introduced a hyperbolic size-density response function for a single species:

$$v(x) = (\alpha + \beta\, x^{\phi})^{-1/\theta} \qquad x, \alpha, \beta, \phi, \theta > 0 \tag{2.12},$$

where x is the density of plants or number of plants in a unit area. The shape parameters ϕ and θ make the response function quite flexible, but only two cases are biologically relevant (Mead 1970, Seber and Wild 1989). If $\phi = \theta = 1$, the cumulative size of plants per unit area, $x\,v(x)$, increases asymptotically with x towards β^{-1} (Fig. 2.6). If $\theta < \phi$, the cumulative size of plants per unit area, $x\,v(x)$, increases towards a maximum at $(\alpha\theta/\beta\,(\phi-\theta))^{1/\phi}$ after which the cumulative size decrease (Fig. 2.6). Generally, at low plant density (for $x \to 0$) $v = \alpha^{-1/\theta}$. The Bleasdale-Nelder model (2.12) may be derived from the Richards growth model (2.7) by assuming an allometric relationship between the size and density, $vx^{\rho} = c$, where c is a constant and $\theta = \delta - 1$, $\phi = (\delta - 1)\rho$ (Seber and Wild 1989).

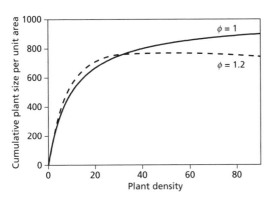

Fig. 2.6 The cumulative plant size per unit area as a function of the shape parameters ϕ in the Bleasdale-Nelder hyperbolic size-density response function (2.12). Full line: $\alpha = 0.01$, $\beta = 0.001$, $\theta = 1$; asymptotic maximum cumulative plant size $\beta^{-1} = 1000$ at the density ∞. Dotted line: $\alpha = 0.01$, $\beta = 0.0005$, $\theta = 1$; maximum cumulative plant size at the density $(\alpha\theta/\beta(\phi-\theta))^{1/\phi} = 46.4$.

It has been found that it is primarily the ratio θ/ϕ in the Bleasdale-Nelder model (2.12) that determines the shape of the plant size – density curve (Mead 1970, Seber and Wild 1989). Therefore a simpler model has been proposed (Watkinson 1980, Vandermeer 1984):

$$v(x) = \frac{v_m}{1 + \beta x^\phi} \qquad (2.13),$$

where θ in (2.12) is set to one. In this simpler model the shape parameters are easier to interpret biologically. v_m ($\approx \alpha^{-1}$) is the size of isolated plants (i.e., for $x \to 0$), β is a measure of competition including its intensity and the area in which it operates, while ϕ is a measure of the rate at which competition decays as a function of distance between plants (Vandermeer 1984).

Example 2.2 Effect of density on *Arabidopsis thaliana*

Arabidopsis thaliana (Brassicaceae) is a small winter annual plant that has been used as a model species in genetic, ecological and evolutionary research. *A. thaliana* (genotype *Nd-1*) was grown in an experimental garden at three densities and the dry weight of the plants was measured after seed setting. It is apparent that density has a negative effect on the dry weight of *A. thaliana* (Fig. 2.7). The dry weight data was fitted to a plant size – density model (2.13) (see Appendix B). The fit was, according to a visual inspection of Fig. 2.7, acceptable, but the maximum likelihood estimates of the parameter values were $\hat{v}_m = 16$ g, $\hat{\beta} = 5200$, and $\hat{\phi} = 1.3$. The estimates of v_m and β are very poor estimates of the size of isolated plants and intensity of competition, respectively. This is due to the well-known phenomenon that empirical models only with the utmost care should be extended outside the domain of the regressed data. If the size of isolated plants was to be estimated correctly then isolated plants should be included in the experiment. Since the parameters v_m and β are correlated, the estimate of β is also poor.

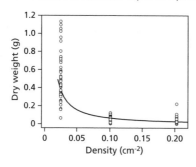

Fig. 2.7 Dry weight of *Arabidopsis thaliana* at three densities, and the fitted line according to plant size – density model (2.13). The variance of dry weight depended highly on density and both the dry weight data and the plant size – density model was Box – Cox transformed ($\lambda_1 = -5$, $\lambda_2 = 1$; see Appendix B).

The individual-based Richards growth model (2.11) may be extended to include growth at different densities by generalising the four parameters in the individual-based growth model (2.11) to functions of density. It is apparent that the average final size, w in (2.11), may be generalised by a plant size – density model, e.g., (2.13), $w(x) = w_m/(1 + \beta x^\phi)$, and assuming that the initial growth rate is independent of density, $(\kappa a)/(\delta-1) = c$, then κ may be generalised to $\kappa(x) = c(\delta(x) - 1)/a(x)$, where c is a constant. The functions $\delta(x)$ and $a(x)$ may be expected to be decreasing and increasing functions of density, respectively, and likely candidate functions may be $\delta(x) = \delta_0 \exp(-\delta_1 x)$ and $a(x) = a_0 + a_1 x$. Generalising the individual-based Richards growth model (2.11) with respect to density require at least four new free parameters, thus the fitting and especially the testing of biological hypotheses will require very good plant growth data and has not been done yet.

Modelling spatial effects

The plant size – density models (2.12) and (2.13) are mean-field models (Durrett and Levin 1994, Levin and Pacala 1997), where it is assumed implicitly that all individuals at a certain density have the same size, and every plant has the same effect of any other plant. The mean-field approach is a sensible place to start in the modelling of plant-plant interactions (Bolker et al. 1997), but it ignores much of what is important about the dynamics of plant communities (Levin and Pacala 1997). In reality, interactions typically are restricted to a subset of the individuals and the likelihood that two plants will interact is most probably a decreasing function of the distance between them (Stoll and Weiner 2000).

As explained above, the negative interactions between neighbouring plants are caused by competition for the same resources. The ways the limiting resources are distributed among the neighbouring plants depend on the type of resource that is limiting for growth, the interacting plant species and their size. Different theoretical models, which describe the distribution of resources, have been proposed. However, the development of theory in this area has been more rapid than the compilation of relevant ecological data and it is difficult to assess the validity of the different models.

The "zone of influence" model is a resource uptake based model, which allows for differences in plant size (Gates and Westcott 1978, Wyszomirski 1983, Hara and Wyszomirski 1994, Weiner et al. 2001). Each plant is assumed surrounded by an imaginary circle (Fig. 2.8), which symbolises the area from which the plant may extract resources. When the plant grows, the radius of the circle expands according to some species-specific rules. After a while the imaginary circles of two neighbouring plants may overlap and

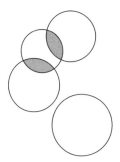

Fig. 2.8 "Zone of influence" model, where an imaginary circle symbolising the area from which each of the four plants may extract resources. The plants share the resources in the overlapping area (grey) according to species-specific rules.

the resources in the overlapping area are shared among the plants according to species-specific rules of sharing resources. For example, the largest plant may get all the resources in the overlapping area (complete size-asymmetric competition), or plants may share the resources in the overlapping area proportionally to their size (perfect size-symmetric competition).

The "zone of influence" model may be generalised by assuming that the influence of the plant within the zone is a decreasing function of the distance from the plant ("field of neighbourhood" model, Berger and Hildenbrandt 2000).

An empirical spatial model called the neighbourhood model assumes that a plant only competes with the neighbours that are positioned within a fitted fixed radius of the plant (Pacala and Silander 1985, Coomes et al. 2002). The competitive effect of the neighbours and the plant itself (self-shading) on the growth of the plant is described by a mean-field plant size – density model (e.g., 2.13). The neighbourhood model has the advantage that it is readily fitted to data (Pacala and Silander 1990). However, in a comparative test between different spatial models on natural populations of *Lasallia pustulata* (Sletvold and Hestmark 1999), the neighbourhood model (number of plants within a circle with fixed radius) was a worse predictor than the distance to the nearest neighbour, which also is a conceptually simpler model.

Given a functional relationship that describes the effect of interplant distances on the competitive interactions, U, such a competitive neighbourhood function may be used to generalise the individual-based Richards growth models (2.11) with respect to the spatial setting:

$$\frac{dv_i(t)}{dt} \begin{cases} \dfrac{\kappa}{1-\delta}\, f(v_i(t),a)\left(\left(\left(\dfrac{1}{w_m}\sum_{j=1}^{n} v_j(t)U(\lambda,r(i,j))\right)^{\delta-1}-1\right)\right) & \delta \neq 1 \\[3mm] \kappa f(v_i(t),a)\left(\log(w_m)-\log\left(\sum_{j=1}^{n} v_j(t)U(\lambda,r(i,j))\right)\right) & \delta = 1 \end{cases}$$

(2.14).

Since the competitive interactions only take place on a local scale, the terms $\sum_{j=1}^{n} v_j(t)$ and $n\,w$ in the saturating terms in (2.11) are

replaced by $\sum_{j=1}^{n} v_j(t)\, U(\lambda, r(i, j))$ and w_m, respectively, where w_m is the final

size of an isolated plant. The competitive neighbourhood function is a two-dimensional density function that describes the negative effect of plant j on plant i as a function of the distance between the two plants, $r(i,j)$ and the scale parameter, λ, which measure the effect of distance on the competitive interactions. However, the effect of distance on the competitive interactions may vary with the mean plant density and the growth stage (Weiner 1984) and λ in (2.14) then becomes a function of more parameters.

The competitive neighbourhood function is analogous to a competition kernel sensu Bolker et al. (2000). One of the likely candidate functions that describes the effect of interplant distances on the competitive effect is the two-dimensional exponential density function (Dieckmann et al. 2000), thus a candidate competitive neighbourhood function may be:

$$U(\lambda, r(i, j)) = \exp(-\lambda\, r(i, j)) \hspace{3cm} (2.15).$$

The fitting of such a spatial explicit individual-based Richards growth model (2.14) requires growth data of individual plants with a known spatial position and the growth model will probably be valuable in the analysis of tree growth, where spatially explicit long-term growth data exists (e.g. Soares and Tomé 1999).

3. Demography

Mortality

In the previous chapter we followed the growth of individual plants to reproductive age (Fig. 1.1), however, some germinated seeds in a synchronous monoculture die before they reach the reproductive age. Plant mortality may be caused by density-independent or density-dependent factors, although, in practice it is difficult to separate the two types of mortality. Harper (1977) gives the following example: "The mortality risk to a seedling from being hit by a raindrop or hailstone might be thought to be density-independent. Presumably the risk of being hit is independent of density but whether a seedling dies after being hit is a function of its size and vigour, both of which are strongly effected by density." Generally, the relative effect of density-dependent factors compared to density-independent factors is expected to increase with plant size in synchronous plant populations because the effect of negative interactions between neighbours will increase with plant size. Thus, density-independent mortality may primarily occur at the seed and seedling stages by, e.g., death during seed dormancy, some forms of seed predation, unfavourable soil conditions, local water availability etc.

Density-dependent mortality or self-thinning in synchronous monocultures show a surprisingly regular pattern: there seems to be a maximum density of surviving plants primarily controlled by the biomass of the plants. In an experiment with buck wheat (*Fagopyrum esculentum*) sown at different densities, the three populations with the highest densities had the same density 63 days after sowing (Fig. 3.1) and the same pattern was found for other species (Yoda et al. 1963).

When the plants in a synchronous monoculture grow, self-thinning will reduce the number of surviving plants if above a maximum density and since the individual growing plants need more and more space and resources, the maximum density decreases with time. For many species it has been found that if the average biomass at a certain time is plotted against the maximum density in a log-log plot, then the relationship is approximately linear with a slope of –3/2 (e.g., Yoda et al. 1963). This so-called "3/2 power law of self-thinning" has been explained theoretically by specific allometric assumptions on plant growth (e.g., Yoda et al. 1963).

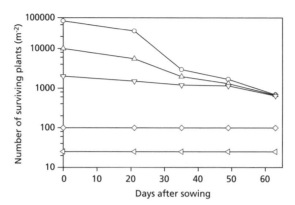

Fig. 3.1 Number of surviving *Fagopyrum esculentum* plants as a function of days after sowing at different sowing densities. Figure after Yoda et al. (1963).

The density-dependent mortality in a synchronous monoculture may be described by assuming that the density of surviving plants, x, is a function of seed density, x_0, (Watkinson 1980):

$$x = \frac{x_0}{1 + m_d x_0} \tag{3.1},$$

where m_d^{-1} is the maximum density after self-thinning at the current average size of the plants.

If density-independent mortality is assumed to occur before density-dependent mortality, then the density of surviving plants after density-dependent mortality may be described by:

$$x = \frac{m_i x_0}{1 + m_d m_i x_0} \tag{3.2},$$

where m_i is the density-independent probability that a seed germinates, gets established, and starts to grow (Firbank and Watkinson 1985). When m_d is large, density-dependent mortality is most important and $x \approx m_d^{-1}$. Opposite, when m_d is small, density-independent mortality is most important and $x \approx m_i x_0$.

Reproduction

The number of seeds produced by a plant, i.e. fecundity, is highly species-specific and mainly determined by the life history and adaptive strategy of the species (Harper 1977, Rees et al. 2001). In a number of empirical studies a trade-off between fecundity and seed size has been demonstrated, whereas seed size and the probability of establishment have been

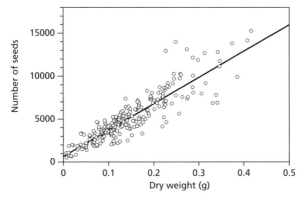

Fig. 3.2 Approximately linear relationship between the biomass of small annuals and the number of seeds produced. The number of seeds produced as a function of dry weight in *Arabidopsis thaliana* (genotype *Nd-1*) (Damgaard and Jensen 2002).

shown to be positively correlated (Rees et al. 2001). Within species, however, the fecundity seems to be positively correlated with plant size.

Annual and some perennial plant species only reproduce once before they die (monocarps), such plants convert all available resources at the time of reproduction into seeds and the fecundity, b, is often found to be a linear function of plant biomass at the time of reproduction (Fig. 3.2).

However other allometric relationships between biomass and fecundity than linearity may be observed, e.g.

$$b(w) = b_0 + b_1 + w^\gamma \tag{3.3},$$

is an often used regression model where γ measures the degree of non-linearity (Schmid et al. 1994). Since a plant necessarily has a certain minimum size before it can reproduce the parameter b_0 may be conditioned to be negative and the problem of fitting such a model to fecundity data, possibly including small plants with zero seeds, has been treated by Schmid et al. (1994).

In many competition experiments, the fecundity has been measured as a function of the density of plants at the reproductive age. Since the relationship between plant size and fecundity is a simple linear function, the average fecundity may be modelled using a mean-field response surface model analogous to a plant size–density response surface model, e.g., the flexible Bleasdale-Nelder response surface model (2.12):

$$b(x) = (\alpha + \beta x^\phi)^{-1/\theta} \tag{3.4},$$

where x is the density of plants at the reproductive age.

Polycarpic perennial plants reproduce usually every year after a juvenile period (although mast seeding occurs in some species e.g. Rees et al. 2002). The perennial plants allocate resources between seed production and investment in structures that increase the survival probability and facilitate growth the following year. There is a weak tendency that the juvenile period is positively correlated with the expected life span of the species (Harper 1977), and in general, the fecundity of polycarpic perennial plants is a function of plant size, which again is a function of age and environment.

Population growth

Combining the above mean-field models of the different density-dependent processes at the different life-stages (Fig. 1.1) the *average* fecundity in a synchronous monocarpic monoculture may be expressed as a function of seed density as $b(w(x(x_0)))$, e.g., using (3.2) and (3.4):

$$b(x(x_0)) = \left(\alpha + \beta \left(\frac{m_i \, x_0}{1 + m_d \, m_i \, x_0}\right)^\phi\right)^{-1/\theta}$$

(3.5).

If the generations are assumed non-overlapping, the density of seeds in the next generation will be a product of the number of surviving plants and their average fecundity:

$$x_0(g + 1) = x(x_0(g)) \, b(x(x_0(g))) =$$

$$\left(\frac{m_i + x_0(g)}{1 + m_d + m_i \, x_0(g)}\right)\left(\alpha + \beta\left(\frac{m_i + x_0(g)}{1 + m_d + m_i \, x_0(g)}\right)^\phi\right)^{-1/\theta}$$

(3.6),

where the time, g, is measured in generations.

In many plant competition studies the fecundity is measured at variable adult plant densities, and the probability of germination, establishment, and reaching reproductive age, $p(x_0)$, is assumed known or estimated in independent experiments. In those cases, the recursive equation (3.6) is most conveniently expressed as the average fecundity of the plants at the reproductive age multiplied with the probability of germination, establishment, and reaching reproductive age in the following generation:

$$x(g + 1) = p(x_0 (g + 1)) \, x(g) \left(\alpha + \beta \, x(g)^\phi\right)^{-1/\theta}$$

(3.7),

where $x_0(g + 1) = x(g) \left(\alpha + \beta \, x(g)^\phi\right)^{-1/\theta}$.

The above model is restricted to monocarpic plants where the ecological success is expected to covary with fecundity, but other fitness measures that covary with the ecological success of a species with a more complicated life-history can easily be adopted. The probability of germination, establishment, and reaching reproductive age in a natural habitat is a critical and variable factor in predicting population growth. Crawley et al. (1993) compared estimates of population growth of transgenic rapeseed (*Brassica napus*) in a natural habitat obtained by either using independent information on seed germination, mortality, and fecundity, or measuring population growth directly by the difference in the number of seedlings from year to year. They obtained qualitatively different estimates of population growth by the two approaches: When the independently obtained information on mortality and fecundity was used, the rapeseed population was predicted to increase in density, whereas the directly obtained estimate of population growth showed that the rapeseed population decreased in density. Later it was observed that the population actually was decreasing until it went extinct (Crawley et al. 2001). Likewise, Stokes et al. (2004) found that variation in the probability of germination and establishment led to a relatively high variation in the population growth rates of *Ulex gallii* and *Ulex minor*.

In a pioneering study (Rees et al. 1996) measured population growth rates, $x(g+1)/x(g)$, directly in a natural habitat for four annual species (*Erophila verna, Cerastium semidecandrum, Myosotis ramosissima* and *Valerianella locusta* subspecies *dunensis*) and modelled the growth rates as a function of density. The population growth of the four annual species was mainly regulated by intraspecific competition (Rees et al. 1996) and could be described by the following relationship, where plant density can be measured at any stage in the life cycle:

$$\frac{x(g+1)}{x(g)} = \lambda(1 + x(g))^{-\psi} \qquad \lambda, \psi > 0 \qquad (3.8),$$

where λ is the density-independent rate of population growth. A similar approach has been taken in the description of the population dynamics of the annual grass *Vulpia ciliata* (Watkinson et al. 2000).

At equilibrium

The above mathematical description of density-dependent population growth leads naturally to the study of the mathematical properties of the dynamical system and in particular to the study of possible equilibria.

Such a mathematical exercise may or may not be relevant for understanding the underlying ecological processes: The successional processes in plant communities are often enduring processes, where many plant species are excluded simply because they are limited by their colonisation ability (Rees et al. 2001). Such plant species only wait for an unlikely immigration event before they will invade the community. Consequently, most plant communities will probably never be at equilibrium. Nevertheless, knowledge on the equilibrium states under certain rather strict assumptions may provide valuable information in predicting the future states of the plant community.

For synchronous monocarpic monocultures the equilibrium density of seeds may be calculated by solving the recursive equation (3.6). The general solution to the recursive equation (3.6), $\hat{x}_0 = x_0(g+1) = x_0(g)$, is quite complex, but two specific cases are worth considering in more detail:

1) When $\phi = \theta = 1$, i.e., the cumulative fecundity per unit area increases asymptotically towards β^{-1}, the nontrivial solution is:

$$\hat{x}_0 = \frac{m_i - \alpha}{m_i \, (\beta + \alpha \, m_d)} \tag{3.9}.$$

The nontrivial equilibrium (3.9) is stable (Appendix D) when $m_i > \alpha$. Note that α^{-1} is the estimated fecundity of an isolated plant ($x \to 0$), thus the probability of reaching the reproductive age has to be higher than the inverse fecundity of an isolated plant for the nontrivial equilibrium (3.9) to be stable. The trivial equilibrium, $\hat{x}_0 = 0$, is stable when $m_i < \alpha$, i.e., the population goes extinct if the probability of reaching the reproductive age is lower than the inverse fecundity of an isolated plant.

2) When only density-independent mortality is occurring, i.e., $m_d = 0 \Leftrightarrow x = m_i \, x_0$, the nontrivial solution to the recursive equation (3.6) is:

$$\hat{x}_0 = \frac{1}{m_i} \left(\frac{m_i^{\,\theta} - \alpha}{\beta} \right)^{1/\phi} \tag{3.10}.$$

The nontrivial equilibrium (3.10) is stable when $m_i > \alpha^{\frac{1}{\theta}}$ and $m_i < \left(\frac{\alpha \phi}{\phi - 2\theta} \right)^{\frac{1}{\theta}}$. The trivial equilibrium is stable when $m_i < \alpha^{\frac{1}{\theta}}$.

When $m_i > \left(\frac{\alpha \phi}{\phi - 2\theta} \right)^{\frac{1}{\theta}}$ the density oscillates around the equilibrium density

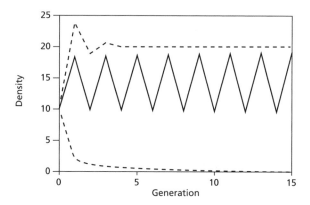

Fig. 3.3 The density of plants at the reproductive age calculated with the recursive equation (3.6) assuming only density-independent mortality, i.e., $m_d = 0 \Leftrightarrow x = m_i\, x_0$.

Solid line: $m_i > \left(\dfrac{\alpha\,\phi}{\phi - 2\theta}\right)^{\frac{1}{\theta}}$; $\alpha = 0.01$, $\beta = 0.001$, $m_i = 0.25$, $\phi = 2.1$, $\theta = 1$, $\hat{x} = 13.6$

Dashed line: $m_i > \alpha^{\frac{1}{\theta}}$; $\alpha = 0.01$, $\beta = 0.001$, $m_i = 0.1$, $\phi = 1.5$, $\theta = 1$, $\hat{x} = 20.1$

Dotted line: $m_i < \alpha^{\frac{1}{\theta}}$; $\alpha = 0.01$, $\beta = 0.001$, $m_i = 0.008$, $\phi = 1.5$, $\theta = 1$, $\hat{x} = 0$

(Fig. 3.3) until, for even larger m_i, when the trajectory becomes chaotic (May and Oster 1976, Hoppensteadt 1982). However, see e.g. Freckleton and Watkinson (2002) for a discussion of the ecological relevance of chaotic systems in describing plant populations.

Alternatively, if mortality is assumed density-independent, $p(x_0) = p = m_i$, the nontrivial equilibrium density of the plants at the reproductive age may be calculated from the recursive equation (3.7):

$$\hat{x} = \left(\frac{p^\theta - \alpha}{\beta}\right)^{1/\phi} \tag{3.11}.$$

The conditions for stability of equilibrium (3.11) correspond to equilibrium (3.10).

The nontrivial equilibrium density calculated from the recursive equation (3.8) of population growth is:

$$\hat{x} = \lambda^{1/\psi} - 1 \tag{3.12},$$

which is stable when $\lambda > 1$, and $\lambda < \left(\dfrac{\psi}{\psi - 2}\right)^\psi$.

Example 3.1   Equilibrium density of *Arabidopsis thaliana*

Arabidopsis thaliana (genotype *Nd-1*) was grown in an experimental garden at three densities and the dry weight of the plants was measured after seed setting (Example 2.2). Using a linear relationship between dry weight and fecundity (Fig. 3.2), the fecundity was estimated for each plant and fitted to the response surface model (3.4). If the probability of establishment is known, then it is possible to calculate the Bayesian posterior distribution (see Appendix C) of the equilibrium density (Fig. 3.4).

The probability of germination, establishment, and reaching reproductive age in a natural habitat is critical for calculating the equilibrium density, however, there exists sparse relevant ecological information for *A. thaliana*. It is assumed that the probability of establishment for *A. thaliana* is relatively low, since the seeds are extraordinarily small and seed size is known to be negatively correlated with the probability of establishment (Rees et al. 2001). Furthermore, since *A. thaliana* is exceptionally plastic and known to be able to survive and reproduce at a large density range, it is assumed that mortality is density-independent.

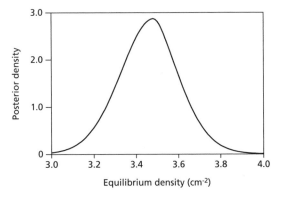

Fig. 3.4 The posterior density distribution of the equilibrium density of *Arabidopsis thaliana*. The shape parameters ϕ and θ were not significantly different from one in a loglikelihood ratio test, and were set to one. Mortality was assumed to be high and density-independent, $p(x_0) = p = 0.001$. The prior distribution was assumed uninformative. Note that the calculated posterior distribution of the equilibrium density is outside the domain of the data (Fig. 2.7) and the result should therefore be verified by growing *A. thaliana* at a density of 3 – 4 cm^{-2}.

Using the stability conditions of the non-trivial equilibrium, it was found that in order to prevent extinction the probability of reaching reproductive age had to be higher than the inverse fecundity of an isolated plant. These calculations assumed implicitly that there was no variation in fecundity or mortality among plants. If there is any random variation in fecundity or mortality in a density-regulated plant population, extinction is expected to occur at some point in time and the mean time to extinction may be calculated under certain assumptions (Lande 1993). The mean time to extinction, that in many cases will be so long that it has no practical consequence, is increasing with equilibrium density and decreasing with variation in fecundity.

Seed dispersal

Seed or vegetative dispersal is a necessary adaptation to a sedentary adult life form. Most plant communities are dynamic with continuous local disturbances followed by a relatively long succession process or may locally go extinct due to demographic or environmental stochasticity (Lande 1993). In such cases plant species have to be able to re-colonise a local area by immigration of seeds or vegetative propagules by the wind or some animal vector (Harper 1977).

It is difficult to obtain good seed dispersal data. It has especially been difficult to quantify the likelihood of rare extraordinary long dispersal events (Bullock and Clarke 2000), which is important for how fast a species may spread due to the establishment of secondary foci (Shigesada and Kawasaki 1997) or an extraordinary fat-tailed dispersal distribution (Kot et al. 1996). When seed dispersal data has been fitted with empirical models either the negative exponential or the inverse power model has traditionally been used since they are conveniently regressed by linear regression models on log-linear or log-log paper, respectively (Bullock and Clarke 2000). In most cases the relatively fat-tailed inverse power model fit seed dispersal data better than the negative exponential model, although a mixture of the two models may give an even better fit (Bullock and Clarke 2000). Unfortunately, these often-used regression models are not probability distributions that sum up to one and therefore do not give a clear picture of the relative importance of a possibly uncensored tail. It would be advantageous to fit two-dimensional probability distributions to the dispersal data, e.g., the exponential density function, $U_{ij}(s) = \lambda^2 \exp(-s\lambda)/2\pi$ with mean dispersal distance $2/\lambda$, the Gaussian density function, $U_{ij}(s) = \exp(-s^2/2\sigma)/2\pi\sigma$ with mean dispersal distance $\sqrt{\pi/2}\sqrt{\sigma}$, or the Bessel density function, $U_{ij}(s) = \lambda^2 \mathrm{BesselK}(0, s\lambda)/2\pi$ with mean dispersal distance $\pi/2\lambda$ (Dieckmann et al. 2000).

An alternative approach to such an empirical fitting of seed dispersal curves is to model the dispersal process using available knowledge on the mechanisms of dispersal. Seed dispersal by means of animals is difficult to model due to the complicated behaviour of many animal vectors and little theoretical work has been done to understand the dispersal pattern of animal-dispersed seeds. In contrast, when seeds are wind-dispersed the physics of the seed and the air is understood in some details. Seed dispersal by wind depends on the seed's weight and volume, as well as the presence of wings or plumes. Wind speed, turbulence and convection currents determine the travelling distance and very small seeds, e.g., many orchid seeds, may travel vast distances. The seed dispersal process may be described by a 3-dimensional diffusion process in the plane from a certain start height above the ground, a, until the seed touch the ground using a measurable downward velocity, g, due to gravitation (Stockmarr 2002). In absence of wind the seed dispersal is isotropic (rotationally symmetric) and the density of dispersal distances, r, may be described by:

$$D(r;a,g) = r\left(\frac{a + g\,a\,\sqrt{a^2 + r^2}}{(a^2 + r^2)^{3/2}}\right)\exp(-g(\sqrt{a^2 + r^2} - a)) \qquad (3.13),$$

which decrease asymptotically as $g\,ar^{-1}\exp(-g\,r)$ that goes even faster to zero than the exponential distribution (Stockmarr 2002) . If the seed's weight to volume ratio is low and gravity may be assumed to be negligible, the density of dispersal distances is:

$$D(r;a) = \frac{r\,a}{(a^2 + r^2)^{3/2}} \qquad (3.14),$$

which is a fat-tailed distribution with infinite mean (Stockmarr 2002). Any knowledge of the mean movement of the air due to wind or turbulence at the time a seed is released (Berkowicz et al. 1986) may be added to the 3-dimensional diffusion process in order to predict the two-dimensional distribution of seed dispersal distances.

Modelling spatial effects

Traditionally the spatial effects in plant ecology have been modelled using lattice models or grid based models, where the spatial effect is modelled by specific rules on how plants positioned in different lattices

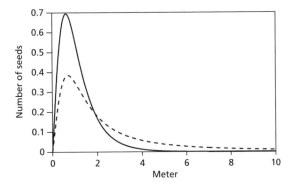

Fig. 3.5 Seed dispersal as modelled by (3.13) and (3.14). The start height above the ground, **a** = 1 meter. Dashed line: **g** = 0 (downward velocity is negligible). Full line: **g** = 1 m s^{-1}. The dispersal distance must be scaled by the diffusion coefficients (Stockmarr 2002).

interact (Dieckmann et al. 2000). Such lattice models may give a very detailed information of the population growth processes, but sometimes it is beneficial to reduce the complexity of the model in order to get more general biological conclusions. Such a reduction in complexity may be obtained by approximating lattice models with comparatively simple recursive equations (Hiebeler 1997, Dieckmann et al. 2000).

Lately there have been some important theoretical advances in the description of the spatial effects in plant ecology in continuous planar space (Dieckmann et al. 2000). The plant demographic processes in continuous planar space has been modelled using fixed rates of fecundity and mortality of identical plants, a dispersal kernel, $D(r)$, and a probability of establishment that is a function of the *local density* at the point u:

$$x(u) = \sum_j U(r(j, u))$$ (3.15),

which is the density weighted according to a competition kernel, $U(r)$, and the location of plants j relative to u (Bolker and Pacala 1999, Bolker et al. 2000). The covariance in local density between any two points u and $u + \Delta u$ is defined as:

$$C(r) = C\big(|\,\Delta u\,|\big) = E\big((x(u) - E(x))\,(x(u + \Delta u) - E(x))\big)$$ (3.16),

where it is assumed that the covariance only depends on the distance between the two points. The *average covariance* over all distances may be defined as the covariance in local density weighted by the spatial scale of the competition kernel and the dispersal kernel:

$$\bar{C} = \int (U * D)(r)\, C(r)\, dr \tag{3.17},$$

where $*$ represents a two-dimensional convolution (Bolker and Pacala 1999, Bolker et al. 2000). It is possible to derive and solve two relatively simple differential equations describing the changes in mean density and average covariance with time assuming that spatial third moment terms may be ignored (Bolker et al. 2000). Such a relatively simple spatial model with known equilibrium conditions may generate important ecological hypotheses, especially when generalised to more species (see next chapter), which may be tested by observations or through manipulated experiments. For example, (Bolker et al. 2000) concluded from numerical studies of the parameter space that, "for realistic parameter values, spatial dynamics lead to even spacing in monocultures".

Unfortunately, the necessary mathematical simplifications, i.e, the assumption of identical plants with fixed rates of mortality and fecundity that immediately starts to reproduce after they have been produced, makes it problematically to fit the above spatial model to plant ecological data using conventional estimation techniques. However, the concept of a local density may be used to generalise more complicated demographic models with respect to spatial effects. It has already been shown how the growth of individual plants may be adequately described by a spatial explicit individual-based Richards growth model (2.14). The mortality during establishment (3.2) may also be described with respect to the local density:

$$x(u) = \frac{m_i\, x_0\,(u)}{1 + m_d\, m_i\, x_0\,(u)} \tag{3.18}.$$

Combining a spatial explicit recruitment mortality model (3.18), a spatial explicit individual-based Richards growth model (2.14), a biomass – seed relationship (3.3), and a seed dispersal kernel e.g. (3.13), which all may be fitted by plant ecological data, may provide the backbone of a spatial demographic simulation model. Such a simulation model may be used to follow the spatial distribution of a specific synchronous monoculture until quasi-equilibrium.

Seed dormancy

Until now plant populations have been assumed to be synchronous, which appear to be a realistic assumption when plants are monocarpic, however, the seeds of many plant species rest in the soil for a variable period of time before they either germinate or die (Harper 1977, Rees and Long 1993). In many habitats the store of seeds buried in the soil (the seed bank) plays an important role in preventing local extinction due to stochastic disturbances or catastrophic events.

Typically, when seed dormancy is investigated experimentally, a recruitment curve is established by following a cohort of seed through time by means of counting the number of emerging seedlings each year. After a number of years the viability of the remaining seeds may or may not be determined. This experimental procedure, although practically the only sensible thing to do, makes it impossible to determine the age-specific germination probability of an individual seed in the seed bank if there is any variability in the probability of mortality and germination among seeds (Vaupel et al. 1979, Rees and Long 1993). Variability in the age-specific mortality and germination probabilities among seeds, which is likely due to the heterogeneous conditions in the soil, makes it impossible to discriminate among alternative models of age-dependent mortality and germination from population-level data. Nevertheless, (Rees and Long 1993) analysed the recruitment curves of 145 annual and perennial plant species with four different models of the age-specific recruitment probability, i.e., the fraction of seeds that germinate this year out of those seeds that become seedlings in the following years (those that do not die). The most simple and often used model of the recruitment probability assume that the recruitment probability is independent of seed age, i.e., the number of seeds which eventually will germinate decrease each year according to a geometric distribution function.

$$R(y; p) = p(1 - p)^y \qquad\qquad (3.19),$$

where $R(y; p)$ is the frequency of seedlings that germinate in year y and p is the recruitment probability. Other models (Rees and Long 1993), including models that takes variability among individual seeds into account, may assume either an increasing, decreasing or non-monotone (first increasing then decreasing) effect of age on the recruitment probability. When the models were fitted to the recruitment curves,

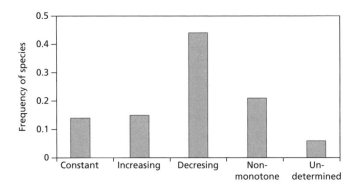

Fig. 3.6 The observed effect of seed age on recruitment probability in 145 plant species.
Figure after Rees and Long (1993).

the recruitment probability was not constant for most species; instead a model where the recruitment probability decreased with age showed the best fit on most recruitment curves (Fig. 3.6). In some cases the recruitment probability increased with age (Fig. 3.6), which may appear as a biologically unrealistic assumption. However, since the recruitment probability is a mixture of the germination and the mortality process, an observed increase in the recruitment probability may be explained by e.g. a constant probability of germination and an increasing probability of dying with seed age (Rees and Long 1993).

Whereas the predicted population growth by the recursive equations (3.6), (3.7), and (3.8) depends on the age-dependent recruitment process, the predicted equilibrium densities (3.9), (3.10), (3.11), and (3.12) are independent of the age structure in the seed bank. At equilibrium, the age structure in the seed bank will also be at equilibrium and a fixed number of seeds will germinate from each age-class in the seed bank each year (Damgaard 1998).

Demographic models of structured populations

For plant species that exhibit seed dormancy or have a perennial life history it is important to consider the effect of age-structure on reproduction and mortality in order to describe population growth processes. The population growth of an age-structured population has traditionally been explored by a Leslie matrix, which is a transition matrix of age-specific survival probabilities and fecundities (Charlesworth 1994, Caswell 2001), or by using the renewal equation in continuous time

(Gurtin and MacCamy 1979, Charlesworth 1994). However, since plants are plastic and have variable growth rates, two plants of the same age will not necessarily have equal survival probability or fecundity. Consequently, the demography of plant populations may be described by a stage-structured matrix model with transition matrices of stage-specific probabilities of moving to another stage and fecundities (Caswell 2001, Stokes et al. 2004), or alternatively, using continuos life history characters by an integral projection model (Easterling et al. 2000).

A matrix population model is a convenient tool in the description of the demographic processes at the existing density and may provide reliable predictions on the immediate future of the plant population. If the transition matrix consists of constant density-independent probabilities and fecundities, the mathematical property of the model is sufficiently simple so that the dominant eigenvalue, which corresponds to the growth rate of the population may be determined (Caswell 2001). The mathematics of density-independent matrix population models has been dealt with in considerable detail by several authors (e.g. Neubert and Caswell 2000, Caswell 2001) and will not be repeated here. However, if the transition probabilities and fecundities are functions of plant density, the mathematical properties of matrix models become somewhat more complicated and has to be calculated using numerical methods (e.g., Stokes et al. 2004).

The complexities of stage-structured population growth with density-dependence are considerably reduced in the important case of an annual plant with a seed bank (Jarry et al. 1995, Freckleton and Watkinson 1998). The life cycle may be reduced to two stages and four processes (Fig 3.7), where relatively simple models, analogous to the models discussed previously, describe the four processes. The four submodels can be rewritten as recursive equations of the densities of seeds in the seed bank and plants at the reproductive age, respectively, and the equilibrium conditions may be found (Jarry et al. 1995, Freckleton and Watkinson 1998).

When fitting a stage-structured demographic model to plant ecological data the quality of the data may depend on the stage. For example, when modelling an annual species with a seed bank, it is simple to get a correct estimate of the density of plants at reproductive age whereas the estimation of the seed density in the seed bank may be difficult resulting in a large observational error. In such cases it may be necessary to model the observational error at each stage (de Valpine and Hastings 2002).

The spatial effects of disturbance and local recruitment in structured plant populations may be modelled in complicated lattice models, which

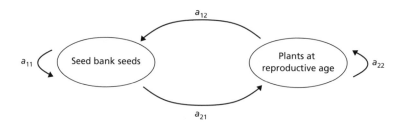

Fig. 3.7 A two-stage life cycle graph for an annual plant with a seed bank. a_{11}: the probability that a seed bank seed remains dormant; a_{12}: number of seeds produced per plant that enter the seed bank; a_{21}: probability of a seed bank seed to survive to reproductive age; a_{22}: number of seeds produced per plant that germinate directly the following year and become adults. Figure after Jarry et al. (1995).

may be approximated with comparatively simple recursive equations. For example, the spatial demography of the perennial bushes *Cytisus scoparius* and *Ulex europaeus* has been described by recursive equations of the frequencies of each spatial site, which is respectively unsuitable for recruitment, open for recruitment, or already occupied by the plant (Rees and Paynter 1997, Rees and Hill 2001).

Long-term demographic data
In order to test various ecological hypotheses, e.g. the effect of density on plant population growth, it is valuable to have plant demographic data where the number or density of plants is recorded during a time series (Rees et al. 1996, Watkinson et al. 2000, Freckleton and Watkinson 2001). Unfortunately, many plant ecological studies have only recorded whether a certain species is present at a certain area and such studies are of limited use in the testing of population ecological hypotheses. Dependent on the species and the local area various methodologies exist in order to quantify the number or density of plants, e.g. counting, pin point analyses, aerial photography etc. (Kjellsson and Simonsen 1994), but in order to get long time series, they require long-term research or monitoring projects.

An interesting possibility of obtaining long-term demographic data is to use fossil pollen to estimate population sizes in the past at different spatial scales. Pollen grains are very robust and are present in high numbers in lake sediments and as the lake sediment gradually builds up during time it is possible to determine when the pollen grain settled in the lake. From such pollen profile data and using species characteristics of pollen production and pollen dispersal, the number of plants within a certain area may be estimated (Gaillard et al. 1998, Clark and Bjørnstad 2004b) (Fig. 3.8).

Alternatively, a demographic history of a species may be determined using a constructed phylogenetic tree based on a sample of DNA sequences. It has been shown that the distribution of coalescence times is a function of whether the population size is constant, decreasing or increasing (Fig. 3.9) and this effect may be explored in a statistical framework to test different demographic models (Emerson et al. 2001, Nordborg and Innan 2002).

Where the fossil pollen data in principle will give a detailed record of the population size with time of several species in a specified area, the phylogenetic approach will only give a very rough picture of population growth or decline. Furthermore, since the historic metapopulation structure as well as the historic growth location of the plant species generally is unknown it is difficult to compare the phylogenies of different species and test effects of interspecific competition.

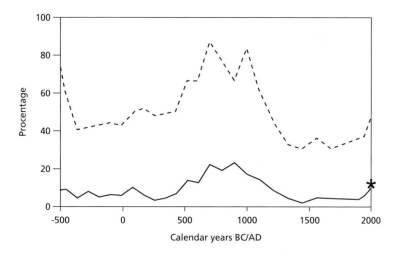

Fig. 3.8 The proportion of tree pollen (dashed line) in a Danish lake and the estimated tree cover in a radius of five kilometres (full line) from 500 BC until today. The present tree cover in a radius of five kilometres is denoted by a star. Figure after Odgaard et al. (2001)

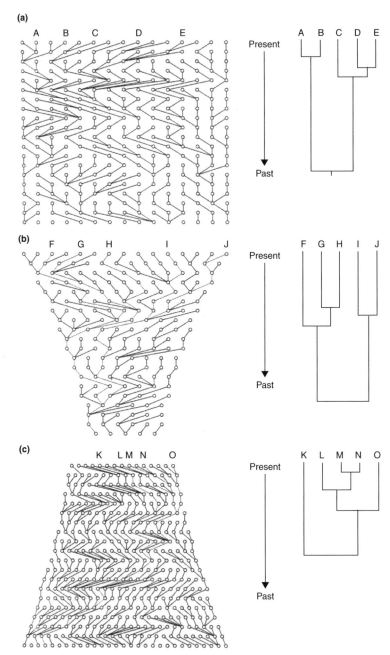

Fig. 3.9 Hypothetical gene trees and corresponding phylogenetic trees from (a) a population of constant size, (b) an exponentially growing population, and (c) an exponentially declining population. In all cases current populations are of equal size (*n* = 15) and five sequences (A-E, F-J, and K-O) were samples from each population. Figure after Emerson et al. (2001).

4. Interspecific competition

Interactions between species

Until now only monocultures have been considered, although, most often populations of different plant species will form plant communities. The individual plants in a natural plant community will typically compete with conspecific plants (intraspecific competition) and with plants belonging to other species (interspecific competition) for the limiting resources (Harper 1977, Goldberg and Barton 1992, Gurevitch et al. 1992).

It is believed that interspecific competition plays an important role in the composition of plant communities, and this has indeed been demonstrated (e.g., Weiher et al. 1998, Silvertown et al. 1999, Gotelli and McCabe 2002). Thus, to understand and possibly to predict the formation of plant communities, the interspecific competitive forces between different plant species have been investigated, often by performing two-species competition experiments (e.g., de Wit 1960, Marshall and Jain 1969, Antonovics and Fowler 1985, Law and Watkinson 1987, Pacala and Silander 1987, Francis and Pyke 1996).

Table 4.1   Different types of interactions between two plant species (after Haskell 1947).

Interaction	Species		Nature of interaction
	A	B	
Competition	-	-	Each species has a negative effect on each other
Parasitism	+	-	Species A exploits species B
Mutualism	+	+	Interaction is favourable to both species
Commensalism	+	0	Species A benefits whereas species B is unaffected
Amensalism	-	0	Species A is inhibited whereas species B is unaffected
Neutralism	0	0	Neither species affect each other

Different plant species have different strategies to obtain their necessary share of the resources in order to grow and reproduce (Grime 2001). Some plant species are of the same functional type, i.e., they compete for the same resources. Other plant species do not compete for the same resources and may have no effect on each other. In some cases the interaction of two species is beneficial for one of the species but has no effect on the other. For example, the early emerging small geophyte *Anemone nemorosa* depends on a dense tree cover of e.g. *Fagus sylvatica* during the summer in order not to be outcompeted by e.g. grasses. In total, two plant species may interact in six qualitatively different ways (Table 4.1).

The nature and strength of interspecific competitive interactions may be investigated by two different approaches: manipulated competition experiments or by censusing coexisting plant populations in a natural plant community. Manipulated plant competition experiments are conducted by measuring size, fecundity, the number of successful descendants, or other measures of ecological success at variable densities and proportions of two or more plant species. There has been an argument in the ecological literature, whether a substitution design (varying species proportions while keeping combined density fixed) or an additive design (increasing the density of one species while keeping the density of the other species fixed) was the best design of a plant competition experiment. This argument is a leftover from the time when plant competition experiments mainly were used to address applied issues in the agricultural sciences. There is no doubt that both types of plant competition designs are equally inferior when plant ecological questions are investigated (e.g. Cousens 1991, Inouye 2001). In the words of Inouye (2001): "The use of substitution and additive designs has largely precluded generating quantitative estimates of the effects of interspecific competition on population dynamics or coexistence, beyond the inference that species do or do not compete." Instead it is recommendable to vary both density and proportion of each species (response surface design) in order to cover a realistic domain of densities and proportions of a natural plant community (Inouye 2001). Hence, the minimum requirement of a two-species competition experiment is three proportions (e.g. 1:0, 1:1, 0:1) at three densities (Fig. 4.1).

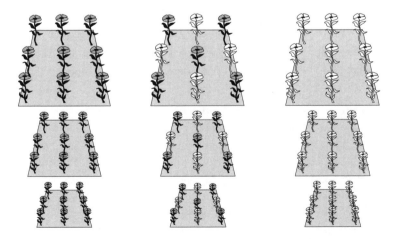

Fig. 4.1 A two-species competition experiment at three proportions (e.g. 1:0, 1:1, 0: 1) and at three densities.

The censusing of coexisting plant populations in a natural plant community (e.g. Rees et al. 1996, Freckleton and Watkinson 2001) has the advantage that it is the actual ecological processes which are studied, where manipulated competition experiments often may be criticised for unrealistic growing conditions. The drawbacks of the censusing methodology is that it is often time consuming and work intensive, and that the domain of the data (densities and proportions of the non-manipulated coexisting species) may be inadequate to make useful ecological predictions.

Modelling interspecific competition

The qualitative descriptive terminology of interaction types in Table 4.1 has become standard, and is readily generalised by the well-known Lotka-Volterra quantitative model of species interaction based on competition coefficients (e.g. Christiansen and Fenchel 1977). The concept of competition coefficients may, at least in principle, be developed from the causal factors: time, habitat and resources of the multidimensional niche concept (MacArthur and Levins 1967, Christiansen and Fenchel 1977). However, usually competition coefficients are thought of as parameters in an empirical competition model, which are estimated using standard statistical methodology (e.g. Marshall and Jain 1969, Harper 1977, Firbank and Watkinson 1985, Law and Watkinson 1987, Pacala and Silander 1990, Francis and Pyke 1996, Rees et al. 1996, Damgaard 1998). This statistical approach has been criticised (Harper 1977, Tilman 1988) for contributing little to the understanding of the underlying mechanisms behind the phenomenon of competition and consequently provide only limited predictive power. In some cases the estimation of competition coefficients may even be directly deceptive of the underlying causes of the interaction, e.g., apparent competition (Holt 1977), when a herbivore tends to eat the most common of two plant species.

As discussed previously, two fundamentally different (or complementary) modelling approaches may be taken in the description of the interactions between plant species: The mechanistic – and the empirical modelling approach. While the mechanistic modelling approach, at least in principle, would respond to the just criticism raised by Harper (1977), Tilman (1988) and others, the complexity and stochasticity of the causal relationships underlying the species interaction in a natural plant community are daunting. The dynamics of plant communities are so complex that only simple heuristic mechanistic modelling is realistic at present. If we want to make use of the available exciting plant ecological data on interspecific competition in a quantitative way, we are forced to make use of relatively simple empirical competition models.

Many manipulated competition experiments have been reported using a set of competition indices introduced by de Wit (1960). Unfortunately, the notion of population changes is not easily incorporated into the de Wit competition model (Inouye and Schaffer 1981) and the model is not readily comparable with the classical Lotka-Volterra competition model. Furthermore, the indices in the de Wit competition model lead to statistical difficulties (Connolly 1986, Skovgaard 1986). Instead, it is advantageous to use a generalised single species competition model, where the effect of the individuals of other species is weighted by competition coefficients. In principle all the models in chapter 2 and 3 might be generalised to multiple species using competition coefficients, but here we will mainly discuss the relatively simple class of mean-field models of two competing species. An often used and flexible mean-field two-species competition model is a generalisation of the hyperbolic size-density response function (2.12) (e.g. Firbank and Watkinson 1985, Law and Watkinson 1987, Damgaard 1998).

$$v_1(x_1, x_2) = \left(\alpha_1 + \beta_1 (x_1 + c_{12} x_2)^{\phi_1} \right)^{-1/\theta_1}$$
$$v_2(x_1, x_2) = \left(\alpha_2 + \beta_2 (c_{21} x_2 + x_2)^{\phi_2} \right)^{-1/\theta_2}$$

(4.1),

where x_i are the densities of plant species i, c_{ij} are the competition coefficients and the other shape parameters are defined as in the single-species case (2.12). The competition coefficient c_{ij} can, analogous to the Lotka-Volterra competition model, be interpreted as the inhibition of species j on species i in units of the inhibition of species i on its own growth. For example, when $c_{ij} = 0$, species j has no effect on the growth of species i; when $c_{ij} = 1$, a plant of species j has the same effect on the growth of species i as a plant of species i; and when $c_{ij} = 2$, one plant of species j has the same effect on the growth of species i as two plants of genotype i. If $c_{ij} < 0$, species j has a positive effect on the growth of species i.

The empirical competition model (4.1) is quite flexible and in many cases the model will be over-parameterised, see the discussion in the single-species case leading to model (2.13). Such a possible over-parameterisation generally decreases the testing power of the model, and it is therefore a standard statistical procedure to test, in this case by a loglikelihood ratio test (Appendix B), whether the model may be reduced by setting e.g. $\theta_1 = \theta_2 = 1$ and $\phi_1 = \phi_2 = 1$.

Example 4.1 Competition between two genotypes of *Arabidopsis thaliana* I

Arabidopsis thaliana is an almost completely self-fertilising winter annual (Abbott and Gomes 1989). The selfing breeding systems means that there is limited genetic exchange between *A. thaliana* genotypes on an ecological time scale (Miyashita et al. 1999) and that the ecological success or fitness of different *A. thaliana* genotypes may be described by a plant species competition model (Ellison et al. 1994). That is, when there is no sexual transmission between individual plants, the ecological success of different genotypes may be modelled as if the genotypes where separate species.

In a manipulated competition experiment two *A. thaliana* genotypes (*Nd-1* and *C24*) were grown in an experimental garden in a design similar to Fig. 4.1, i.e., three proportions (1:0, 1:1, 0:1) at three densities (0.025 cm^{-2}, 0.101 cm^{-2}, 0.203 cm^{-2}), and the dry weights were measured after seed setting.

The dry weight data was fitted to competition model (4.1) and after the model was reduced ($\theta_1 = \theta_2 = \phi_1 = \phi_2 = 1$, loglikelihood ratio test with four degrees of freedom, P = 0.36), the posterior distribution of the competition coefficients were calculated assuming an uninformative prior (Fig. 4.2). Based on the 95% credibility intervals (Appendix C) genotype *Nd-1* has a significantly higher negative effect on *C24* than vice versa.

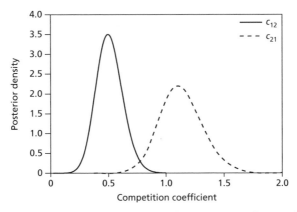

Fig. 4.2 The posterior density distribution of competition coefficients of two *Arabidopsis thaliana* genotypes (1: *C24* and 2: *Nd-1*). The 95 % credibility intervals for c_{12} is {0.30 – 0.75} and for c_{21} is {0.80 – 1.52}.

Analogous to the single-species case in Chapter 3, discrete recursive equations of the densities of competing synchronous monocarpic plant populations may be formulated. However, in order to simplify the calculations it will in the following be assumed that the probability of germination, establishment and reaching reproductive age is independent of the seed densities of the competing species and constant, i.e. $p_i(x_{i,0}, x_{j,0}) = p_i$, and the recursive equations will be:

$$x_1(g+1) = p_1\, x_1(g)\left(\alpha_1 + \beta_1(x_1(g) + c_{12}\, x_2(g))^{\phi_1}\right)^{-1/\theta_1}$$
$$x_2(g+1) = p_2\, x_2(g)\left(\alpha_2 + \beta_2(c_{21}x_1(g) + x_2(g))^{\phi_2}\right)^{-1/\theta_2}$$

(4.2),

where the hyperbolic terms are measures of the average fecundity per plant (Hassell and Comins 1976, Firbank and Watkinson 1985, Damgaard 1998). The parameters in competition model (4.1) may be estimated from competition experiments and the population growth of the two species may be predicted using (4.2). To account for the possible effect of density-dependent mortality and if density-independent mortality may be assumed to occur before any density-dependent mortality (see Chapter 3), then plant densities should be censused before the onset of density-dependent mortality. The plants that die due to density-dependent mortality before they are able to reproduce will then be recorded as having a fecundity of zero.

Analogous to the single-species case, knowledge on the equilibrium densities may provide valuable information in predicting the future states of the plant community. The recursive equations (4.2) may be solved, $\hat{x}_i = x_i(g + 1) = x_i(g)$, with the equilibria (Damgaard 1998):

$$\hat{x}_1 = 0; \qquad \hat{x}_2 = 0 \qquad\qquad (4.3a),$$

$$\hat{x}_1 = u_1; \qquad \hat{x}_2 = 0 \qquad\qquad (4.3b),$$

$$\hat{x}_1 = 0; \qquad \hat{x}_2 = u_2 \qquad\qquad (4.3c),$$

$$\hat{x}_1 = \frac{u_1 - c_{21}u_2}{1 - c_{12}c_{21}}; \quad \hat{x}_2 = \frac{u_2 - c_{12}u_1}{1 - c_{12}c_{21}} \qquad (4.3d),$$

where $u_i = (\beta_i^{-1}(p_i^{\theta_i} - \alpha_i))^{1/\phi_i}$.

In order to determine when the nontrivial equilibrium (4.3d) degenerated to either equilibrium (4.3b) or equilibrium (4.3c); equilibrium (4.3d) was solved for c_{ij} after setting $\hat{x}_1 = \hat{x}_2 = 0$ ($c_{12}c_{21} \neq 1$) and the following roots were obtained:

$$\breve{c}_{12} = u_1/u_2 \; ; \quad \breve{c}_{21} = u_2/u_1 \tag{4.4}.$$

The Jacobian matrix (see Appendix D) of recursive equation (4.2) at the nontrivial equilibrium (4.3d) has two complicated eigenvalues $\{\lambda_1, \lambda_2\}$. The inequalities $\lambda_1 < 1$ and $\lambda_2 < 1$ can be solved and both has the solution:

$$c_{12} < \breve{c}_{12}; \quad c_{21} < \breve{c}_{21} \tag{4.5}.$$

The solutions to the inequalities $\lambda_1 > -1$ and $\lambda_2 > -1$ are more complicated and a numerical investigation of the eigenvalues is necessary to determine the stability of a specific equilibrium. Generally, if the curves of the cumulative plant sizes are not to concave (see Fig. 2.6), then the two species will coexist at equilibrium at the equilibrium densities (4.3d) if both species are able to persist when alone (see stability conditions for equilibrium (3.10)) and inequalities (4.5) are fulfilled. In the important case, when $\phi_i = \theta_i = 1$ and $c_{12} > 0$, $c_{21} < 1/c_{12}$, it can be showed that if both $\lambda_1 < 1$ and $\lambda_2 < 1$ then $\lambda_1 > -1$ and $\lambda_2 > -1$ (Damgaard 2004a). When the curves of the cumulative plant sizes becomes sufficiently concave then the nontrivial equilibrium (4.3d) becomes a saddle point and the trajectory of the densities of the two species bifurcates into periodic coexisting densities. For even more concave curves of the cumulative plant sizes the dynamics become chaotic and the two species coexist at densities in the form of a strange attractor (Damgaard 2004a).

Analogous to the continuous Lotka-Volterra competition model, there are four different ecological scenarios when two species compete: coexistence, species 1 will outcompete species 2, species 2 will outcompete species 1, and either species may outcompete the other depending on the initial conditions. In the last case of indeterminate competition the rarer of the two species will generally be outcompeted (but see Hofbauer et al. 2004). These four ecological scenarios may be characterised by a set of inequalities of the competition coefficients similar to (4.5) (Table 4.2).

Table 4.2 The four different ecological scenarios when two species compete (Damgaard 1998).

Ecological scenario	Condition	Equilibrium
Coexistence	$c_{12} < \check{c}_{12}; \; c_{21} < \check{c}_{21}$	4.3d
Species 1 will win	$c_{12} < \check{c}_{12}; \; c_{21} > \check{c}_{21}$	4.3b
Species 2 will win	$c_{12} > \check{c}_{12}; \; c_{21} < \check{c}_{21}$	4.3c
Either species 1 or species 2 will win	$c_{12} > \check{c}_{12}; \; c_{21} > \check{c}_{21}$	4.3b or 4.3c

In some applied ecological questions, e.g., risk assessment of genetically modified plants (Damgaard 2002) and the management of natural habitats, it is desirable to be able to predict which of the four different ecological scenarios is most likely. If the competition model (4.2) is reparameterised so that $\check{c}_{ij} = c_{ij} + \delta_{ij}$, then the signs of δ_{ij} will discriminate between the four ecological scenarios (Fig. 4.3). The four different ecological scenarios may be considered as four complementary hypotheses and the Bayesian posterior probabilities of each hypothesis may be calculated from a manipulated competition experiment, a known and density-independent probability of reaching reproductive age, and a prior distribution of the four hypotheses (Damgaard 1998).

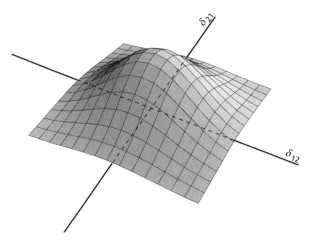

Fig 4.3 The joint posterior distribution of the two competition coefficients ($c_{ij} = \check{c}_{ij} + \delta_{ij}$) calculated from a hypothetical two-species competition experiment and the recursive equations (4.2) (Damgaard 1998). The volume under the "posterior surface" in the quadrate $\delta_{12} < 0$ and $\delta_{21} < 0$ is equal to the probability that the two species will coexist at equilibrium. Likewise, the volumes in the quadrate $\delta_{12} < 0$ and $\delta_{21} > 0$, or $\delta_{12} > 0$ and $\delta_{21} < 0$, is equal to the probability that either species one, or two, respectively, will outcompete the other species, and the volume in the quadrate $\delta_{12} > 0$ and $\delta_{21} > 0$ is equal to the probability that either species will outcompete the other species.

Example 4.2 Competition between two genotypes of *Arabidopsis thaliana* II

Assume that the two genotypes of *Arabidopsis thaliana* (*Nd-1* and *C24*) investigated in example 4.1 are the two only genotypes in the *A. thaliana* population and that the density of *A. thaliana* is not controlled by other species. Similar to example 3.1, assume that mortality is high and density-independent, and furthermore that both genotypes have the same probability of reaching reproductive age.

The fecundity was estimated from the dry weight data by linear regressions. The two genotypes differed significantly in the way they converted biomass at the end of the growing season into fecundity; genotype *C24* produced relatively more seeds per biomass (Damgaard and Jensen 2002). The Bayesian posterior probabilities of each ecological scenario was calculated assuming that the different ecological scenarios were equally probable (uninformative prior distribution) (Table 4.3). Depending on the probability of germination, establishment and reaching reproductive age either "coexistence of the two genotypes" or "genotype *C24* would outcompete *Nd-1*" was the predicted most likely long-term ecological scenario.

Table 4.3 Predicted probabilities of the four different ecological scenarios when two genotypes of *Arabidopsis thaliana* (*Nd-1* and *C24*) compete against each other at different probabilities of germination, establishment, and reaching reproductive age.

Ecological scenario	$p_1=p_2=0.005$	$p_1=p_2=0.001$	$p_1=p_2=0.0005$
Coexistence	0.34	0.57	0.71
Genotype *Nd-1* will win	0	0.001	0.003
Genotype *C24* will win	0.66	0.43	0.28
Either genotype will win	0	0	0.001

In the single-species case, it was discussed that the probability of germination, establishment and reaching reproductive age in a natural habitat is a critical and variable factor in predicting population growth and equilibrium densities (Crawley et al. 1993, Stokes et al. 2004). However, as suggested in Table 4.3, the predicted probabilities of the four different ecological scenarios when two species compete against each other are less sensitive to variation in the absolute value of the establishment probabilities. It is the relative differences between the establishment probabilities, the fecundities and the competitive abilities of the two

Example 4.3 Competition between *Avena fatua* and *Avena barbata*

Assume that the densities of *Avena fatua* and *Avena barbata* are controlled by each other through competitive interactions and not by other species, and furthermore that the probabilities of reaching reproductive age are density-independent and known.

The fecundities of the two *Avena* species were estimated in a competition experiment of five proportions at six densities (Marshall and Jain 1969). The Bayesian posterior probabilities of each ecological scenario were calculated assuming that each of the ecological scenarios was equally likely to occur (uninformative prior distribution) (Table 4.4).

Ecological scenario	$p_1 = p_2 = 0.25$
Coexistence	0.186
A. fatua will win	0.671
A barbata will win	0.001
Either species will win	0.142

Table 4.4 Predicted probabilities of the four different ecological scenarios when *Avena fatua* and *Avena barbata* compete against each other (Damgaard 1998).

The species distribution of *Avena fatua* and *Avena barbata* was observed in a number of natural plant communities in two regions in California (Marshall and Jain 1969). Interestingly, the predicted probabilities correspond with the observed distribution of the two plant species in the Mediterranean warm summer region, whereas the observed competitive interactions do not explain the observed plant distribution in the Mediterranean cool summer region (Fig. 4.4).

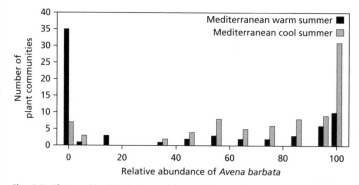

Fig. 4.4 The species distribution of *Avena fatua* and *Avena barbata* in two regions in California. Figure after Marshall and Jain (1969).

species, which mainly determine the fate of the competitive interaction (Damgaard 1998). Consequently, the predictions based on the competitive interactions between two species may therefore be less variable from year to year than estimating the population growth of each species separately.

Similar to the single-species case the predicted equilibrium state of the plant community is independent of the age structures in the seed bank. At equilibrium, the age structures in the seed bank will also be at equilibrium and a fixed number of seeds will germinate from each age-class and species in the seed bank each year (Damgaard 1998).

Ecological predictions will always have to be taken with some scepticism; it is an open question how well the predicted probabilities of the different ecological scenarios obtained from a short-term competition experiment will reflect actual long-term ecological processes in a natural plant community (Kareiva et al. 1996). The realism will to a large extent be determined by the design of the competition experiment, since the predicted probability will never be better than the competition experiment. If a key factor determining the competitive output between two plant species is not included in the experiment, then the predicted probabilities will most likely be erratic and misleading.

Manipulated experiments necessarily reduce the ecological complexity, and extrapolation from such experiments to long-term ecological scenarios can only be made with uncertainty. The predicted probabilities of the different ecological scenarios depend critically on a number of abiotic and biotic factors that for practical reasons often are kept fixed in manipulated experiments. The competitive ability may vary with the physical environment, e.g., temperature, nutrition and water availability (Clauss and Aarssen 1994). Species may also perform differently in competition with other plant species (Abrams 1996, Rees et al. 1996). Nevertheless, when confronted with a complex system, it is common scientific methodology to reduce the complexity of the system by ignoring some processes, to obtain manageable information, which is expected to be relevant in understanding the full system. Furthermore, quantitative ecological predictions are in demand in applied ecology, e.g., risk assessment of genetically modified plants and the management of natural habitats, and more generally to advance the scientific field of plant ecology (Keddy 1990, Cousens 2001).

In the above discussion of the equilibrium states of two competing plant species, it is implicitly assumed that the environment where the plant species are competing is approximately constant or slowly changing on an ecological time scale. However, in some ecosystems, e.g., competing algae species, the species composition may have a large effect on the environment; in the algae case the amount of abiotic essential resources is altered. In such ecological systems more complicated dynamics, analogous to a predator-prey system of interaction, may be a more relevant description of the system (Huisman and Weissing 2001).

Modelling spatial effects

Often plant species are non-randomly distributed among each other (Dieckmann et al. 2000). The non-randomness may be due to historic establishment events of a primarily stochastic nature or as an effect of competitive interactions in previous generations, and the spatial processes of plants have theoretically been shown to affect the structuring of plant communities (Bolker and Pacala 1999). There has been a recent trend within the scientific community to explain various general ecological phenomenons by spatial effects and new hypotheses on the role of space in ecological processes have been developed (Tilman and Kareiva 1997, Dieckmann et al. 2000). Unfortunately, the theoretical investigation of the often-complicated spatial effects shows a history of being prone to erroneous interpretations (Pacala and Levin 1997). For example, in an often-cited paper Tilman (1994) showed that an arbitrarily large number of competing species can coexist in a spatially structured habitat, but later this effect has been shown to be due to an overly simplified competition model rather than to spatial structure (Adler and Mosquera 2000).

It is possible to generalise the empirical models developed for the single-species case using competition coefficients and test different ecological hypotheses with plant ecological data. However, only a few empirical studies have actually used such models to explain the effect of space on interspecific competition (Pacala and Silander 1990, Coomes et al. 2002) and these studies have not demonstrated a significant effect of space on interspecific competition between two pairs of annual species.

The importance of spatial effects in interspecific competition and the structuring of plant communities have until now mainly been examined by heuristic models. Analogous to the single-species case (equations 3.16 and 3.17), the spatial covariance of two species ($C_{11}(r)$, $C_{22}(r)$, and $C_{12}(r)$) as a function of the distance, r, at which it is measured, and the average spatial covariance

$$\bar{C}_{ij} = \int (U_{ij} * D_i)(r) C_{ij}(r)\, dr \tag{4.6},$$

(Bolker and Pacala 1999, Bolker et al. 2000), may be defined in the two-species case. Comparable to the single-species case, a set of differential equations describing the changes in mean densities and average covariances in a spatial explicit model of plants with a simplified life history may be formulated and the invasion criteria of an invading species may be calculated. In this way the invasion criteria of different spatial plant strategies have been investigated (Bolker and Pacala 1999).

In the discrete hyperbolic competition model (equation 4.1) the plants are implicitly assumed to be randomly dispersed, however, the notion of the average spatial covariance may be integrated into the competition model (Bolker and Pacala 1999, Damgaard 2004b):

$$v_1(x_1,x_2) = \left(\alpha_1 + \beta_1\,(x_1 + \bar{C}_{11}/x_1 + c_{12}(x_2 + \bar{C}_{12}/x_2))^{\varphi_1}\right)^{-1/\theta_1}$$
$$v_2(x_1,x_2) = \left(\alpha_2 + \beta_2\,(c_{21}(x_1 + \bar{C}_{12}/x_1) + x_2 + \bar{C}_{22}/x_2)^{\varphi_2}\right)^{-1/\theta_2}$$

(4.7),

where $x_i + \bar{C}_{ii}/x_i$ is the average density of species i in the vicinity of species i, and $x_i + \bar{C}_{ii}/x_i$ is the average density of species i in the vicinity of species j (Bolker and Pacala 1999). Assuming that the average spatial covariance at equilibrium is known, it is possible to calculate the predicted probabilities of the different ecological scenarios as explained previously.

A special issue of spatial covariance is that many manipulated plant competition experiments for practical purposes often are arranged in a non-random spatial design, i.e. a grid design, a row design, or a honeycomb design. Non-random spatial designs have consequences for the analysis of competition experiments that need to be clarified in order to interpret the results (Mead 1967, Fortin and Gurevitch 2001, Stoll and Prati 2001, Damgaard 2004b). A regular spatial design in manipulated competition experiments decreases the variation in size and weight among conspecific plants since the species composition and density of the neighbourhood in most designs are held constant and less variable than in a random spatial design. Such a decrease in variation increases the likelihood of detecting a difference in the competitive ability of different plant species, which in many cases may be a motivating factor for choosing a non-random design. It could be argued that a reduction in experimental variation in many cases would be beneficial since the objective of many experiments is to detect mean differences rather than describing the variation. However, for a deeper understanding of the competitive forces and in order to predict different ecological scenarios, it is necessary to estimate different parameters of interests in a competition model. The problem with using a non-random design is that the data from plant competition experiments usually are analysed in mean-field competition models, which implicitly assume that plants interact

Example 4.4   Competition between two genotypes of *Arabidopsis thaliana* III

The two genotypes of *Arabidopsis thaliana* (*Nd-1* and *C24*) investigated in examples 4.1 and 4.2 were grown in a lattice grid design and in the mixed treatment the genotypes were arranged in a chessboard pattern. The spatial covariance can be calculated from such a regular spatial pattern by using the probability of site occupancy for each plant species and the conditional probabilities of two plants being the same plant species and different plants species (Damgaard 2004b).

There is no dispersal in the competition experiment and the average spatial covariance may be calculated with respect to a known competition kernel (equation 4.6). However, since there is no prior knowledge on the spatial scale of the interaction distance and the functional shape of the competition kernel a sensitivity analysis of different spatial scales and functional shapes was made (Table 4.5). The parameters of interest and especially the predicted most likely ecological scenario depended strongly on the mean interaction distance of the competition kernel, whereas the functional shape of the competition kernel was less important (Table 4.5). Since the actual competition kernel among *A. thaliana* plants is unknown it is difficult to interpret the consequences of the results in Table 4.5, except that neglecting the effect of spatial covariance in non-randomly designed competition experiments may affect the inferred conclusions (Damgaard 2004b).

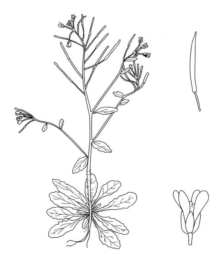

Table 4.5 Estimation of parameters of interest from a competition experiment between two A. thaliana genotypes (1: C24 and 2: Nd-1) under field conditions with and without including spatial covariance (mean-field) using equation (4.7). The competition kernel at various mean interaction distances was assumed to be either a two-dimensional exponential -, Gaussian -, or Bessel distribution with the dimension of the scale parameter in cm. The competition kernel is expected to differ among species, i.e., $U_{11} \neq U_{22} \neq U_{12} \neq U_{21}$, but since the two A. thaliana genotypes belong to the same species it is here assumed that $U = U_{11} = U_{22} = U_{12} = U_{21}$.

There is no prior knowledge on the equilibrium spatial distribution of the two A. thaliana genotypes and therefore it is assumed that the genotypes at equilibrium are randomly dispersed, but it is possible to generalise this assumption. The establishing probabilities for both genotypes were set to 0.001 (Damgaard 2004b). Note that, based on the same set of experimental data, the predicted outcome of competition between two Arabidopsis thaliana genotypes can vary between (almost certain) coexistence and (almost certain) exclusion of one genotype, depending on the assumptions made about the spatial scale of competition between the plants. The predicted outcome of competition depended strongly on the mean interaction distance of the competition kernel, whereas the functional shape of the competition kernel was less important.

	Max. likelihood:		Predicted probabilities of ecological scenarios:			
	c_{12}	c_{21}	Co-existence	Only C24	Only Nd-1	Either
Mean interaction distance = 1						
exponential $\lambda = 2$	1.240	0.491	0.438	0.562	0	0
Gaussian $\delta = 2/\pi$	1.107	0.498	0.528	0.471	0.0002	0.0004
Bessel $\lambda = \pi/2$	1.094	0.522	0.461	0.539	0	0
Mean interaction distance = 2						
exponential $\lambda = 1$	1.104	0.602	0.175	0.824	0	0
Gaussian $\delta = 8/\pi$	1.102	0.656	0.148	0.852	0	0
Bessel $\lambda = \pi/4$	1.199	0.667	0.213	0.787	0.0003	0
Mean interaction distance = 20						
exponential $\lambda = 0.1$	1.082	0.401	0.678	0.320	0.001	0.001
Gaussian $\delta = 800/\pi$	1.078	0.425	0.800	0.200	0.001	0
Bessel $\lambda = \pi/40$	1.155	0.460	0.673	0.325	0.001	0.0003
Mean interaction distance = ∞						
mean-field	1.114	0.501	0.572	0.427	0.0007	0.0002

with other plants in a random way. It is therefore suggested that plants in manipulated competition experiments should be placed randomly, so that the design of the competition experiment is in agreement with the model used in the analysis (e.g. Damgaard 2004b).

Environmental gradients

The abiotic and biotic environment in which most plant interactions take place is highly variable across space and time. It has long been recognised that temporal and spatial variation in the environment is a major force that may influence the outcome of interspecific plant competitive interactions and plant community structures (Tilman 1988, Grime 2001). However, this important notion has only recently begun to influence the theoretical population dynamic models of species interaction (Chesson 2003).

The nature of the spatial and temporal environmental variation also varies and different modelling approaches may be used depending on the studied environmental variation (e.g. Rees et al. 1996, Bolker 2003), but here we will focus on the modelling of a spatial environmental gradient. An environmental gradient may be defined as a set of locations that vary with respect to one or more environmental factors and where the environment of each location is approximately constant or slowly changing on an ecological time scale. Different theoretical hypotheses on the expected effect of various environmental gradients have been developed and investigated in a number of empirical studies. For an introduction to the various hypotheses and the empirical work see e.g. Tilman (1988), Greiner La Peyre et al. (2001), and Grime (2001).

Often studied environmental gradients are specific abiotic stress factors like water availability, salinity, nitrogen availability, heavy metal concentration etc., but also the effects of general productivity (a summary indicator of plant growth) has been investigated. However, in some cases biotic factors may be assumed to be sufficiently constant on an ecological time scale to be modelled as an environmental gradient, e.g., the effect of shading trees on herbaceous plants and the effect of general herbivores and pathogens (Damgaard 2003b).

The growth and competitive interactions of two plant species along an environmental gradient may be adequately described by a generalisation of the discrete hyperbolic competition model (4.1), where it is assumed that the plants are effected by the level (h) of a specific environmental factor,

$$v_1(x_1, x_2, h) = \left(f_{\alpha_1} + f_{\beta_1} (x_1 + f_{c_{12}} x_2)^{\phi_1} \right)^{-1/\theta_1}$$
$$v_2(x_1, x_2, h) = \left(f_{\alpha_2} + f_{\beta_2} (f_{c_{21}} x_1 + x_2)^{\phi_2} \right)^{-1/\theta_2}$$

(4.8),

by some functions $f_z = f_z(z_0, h)$, where $h \geq 0$ and $f_z(z_0, 0) = z_0$ (Damgaard 2003b). The functions f_z that capture the effects of the environmental factor are of course generally unknown and depend on the environmental factor. There may exist prior knowledge, which will aid in choosing the right functional relationship. Alternatively, if competition experiments are made at several different levels of the specific stress, then the functional relationship may be chosen by likelihood ratio tests or by the use of e.g. the Akaike information criterion of different candidate functions. The exponentially decreasing function, sigmoid dose – response function, or the linear function will in many cases be natural candidate response functions, and these response functions may be used interchangeably in the outlined methodology. For simplicity, it is here assumed that within a certain limited range of the stress level, the stress affects the competitive interactions and the reproductive fitness of the susceptible plant species linearly, i.e.,

$$v_1(x_1, x_2, h) = \left(\alpha_{1,0} + \alpha_{1,1}h + (\beta_{1,0} + \beta_{1,1}h)(x_1 + (c_{12,0} + c_{12,1}h)x_2)^{\phi_1}\right)^{-1/\theta_1}$$
$$v_2(x_1, x_2, h) = \left(\alpha_{2,0} + \alpha_{2,1}h + (\beta_{2,0} + \beta_{2,1}h)((c_{21,0} + c_{21,1}h)x_1 + x_2)^{\phi_2}\right)^{-1/\theta_2}$$

(4.9).

Assuming a density-independent and constant probability of reaching reproductive age the discrete recursive equations of the densities of two competing synchronous monocarpic plant populations is:

$$x_1(g+1) = p_1 x_1(g) v_1(x_1(g), x_2(g), h)$$
$$x_2(g+1) = p_2 x_2(g) v_2(x_1(g), x_2(g), h)$$

(4.10),

where $v_i(x_1(g), x_2(g), h)$ is a measure of the average fecundity per plant. The recursive equations have the following equilibrium solutions:

$$\hat{x}_1 = 0; \qquad\qquad \hat{x}_2 = 0 \qquad\qquad\qquad (4.11a)$$

$$\hat{x}_1 = u_1; \qquad\qquad \hat{x}_2 = 0 \qquad\qquad\qquad (4.11b)$$

$$\hat{x}_1 = 0; \qquad\qquad \hat{x}_2 = v \qquad\qquad\qquad (4.11c)$$

$$\hat{x}_1 = \frac{u - f_{c_{12}}(h)v}{1 - f_{c_{12}}(h)f_{c_{21}}(h)}; \qquad \hat{x}_2 = \frac{v - f_{c_{21}}(h)u}{1 - f_{c_{12}}(h)f_{c_{21}}(h)} \qquad (4.11d)$$

where $u = \left(\frac{p_1^{\theta_1} - (\alpha_{1,0} + \alpha_{1,1}h)}{(\beta_{1,0} + \beta_{1,1}h)}\right)^{\phi_1^{-1}}$, $v = \left(\frac{p_2^{\theta_2} - (\alpha_{2,0} + \alpha_{2,1}h)}{(\beta_{2,0} + \beta_{2,1}h)}\right)^{\phi_2^{-1}}$, and $f_{cij}(h) = c_{ij,0} + c_{ij,1}h$

(Damgaard 2003b). A local linear stability analysis of the recursive equation (4.10) at the different equilibria (4.11 a-d) can be made. The eigenvalues are too complicated to be of general use, but they can be calculated numerically in specific cases, so that the stability properties of the different equilibria may be known.

If species 1 is more tolerant to the environmental factor than species 2, then the condition when species 1 will outcompete species 2 can be found by rearranging the equations after setting the nontrivial equilibrium (4.11d) equal to (4.11b) (Damgaard 2003b):

$$f_{c_{21}}(h) > \frac{v}{u} \qquad (4.12).$$

Plant – herbivore and plant – pathogen interactions

Above the effect of a relatively constant environmental factor on plant competition was discussed and it was argued that general herbivores or pathogens for modelling purposes might be considered to be relatively constant environmental factors. However, if the herbivore or pathogen (in the following called a parasite) has one or more of the investigated competing plant species as the most important host plant, then the density of the parasite is controlled by the density of the host plants. That is, any changes in host plant size or density due to either competition, herbivory or disease will affect the density of the parasite.

The population growth of two competing plant species and a specific parasite may be modelled by recursive equations:

$$x_1(g + 1) = p_1\, x_1(g)\, v_1(x_1(g), x_2(g), h(g))$$

$$x_2(g + 1) = p_2\, x_2(g)\, v_2(x_1(g), x_2(g), h(g)) \qquad (4.13),$$

$$h(g + 1) = f(x_1(g), x_2(g), h(g))$$

where $f(x_1(g), x_2(g), h(g))$ describes the density of the parasite as a function of the densities of the two plant species and the parasite at the previous plant generation. Depending on the parasite life history and especially the generation time, likely candidate functions of $f(x_1(g), x_2(g), h(g))$ are various empirically fitted standard discrete population growth models, a Nicholson-Bailey type of model, and others (Hudson and Greenman 1998). Some parasite species tend to concentrate on a single food source and in some cases a population of parasites will switch between the two competing species as the preferred food source depending on the plant densities. Such a system of an optimal foraging parasite stabilises the system so that local extinction events will become less likely (Krivan 1996, Krivan and Sidker 1999). Most models describing the population growth of two competing plant species and a specific parasite will face mathematical problems; in many cases only numerical methods will be available to find possible stable equilibria and this complicates the calculations needed to make ecological predictions using Bayesian statistics.

The complicated subject of how another trophic level affect the interaction of species is a classic ecological question and has been studied extensively both theoretically and empirically. It is out of the scope of this monograph to examine the subject in any detail. Here it will suffice to mention that depending on how the parasite affect the competing plant species, the parasite may either enhance or disrupt the likelihood that the two plant species coexist at equilibrium (Yan 1996, Hudson and Greenman 1998).

Plant strategies and plant community structure

Most plant communities are dynamic with continuous local disturbances followed by a relatively long succession process, where plant species have to be able to re-colonise a local area (Rees et al. 2001). Early-successional plant species typically have a series of correlated traits, including high fecundity, long-distance dispersal, rapid growth when resources are abundant and slow growth and low survivorship when resources are scarce. Late-successional species usually have the opposite traits, including relatively low fecundity, short dispersal distances, slow growth, and an ability to grow, survive, and compete under resource-poor conditions (Grime 2001, Rees et al. 2001). It has been hypothesised that much of the plant species diversity among plant communities is controlled by a trade-off between the ability to colonise new habitats and the ability to compete for resources. In plant communities there tend to be considerably more small seeded plant species than large seeded plant species suggesting, in concert with some seed addition experiments, that many plant species are limited by their ability to colonise new habitats (Rees et al. 2001).

As discussed above, the invasion criteria of different spatial plant strategies of plants with a simplified life history may be investigated by describing the changes in mean densities and average covariances in a spatial explicit model (Bolker and Pacala 1999). If plant species are divided into long-distance (globally) dispersing species and short-distance (locally) dispersing species, then only three different spatial strategies may invade under the assumption that the resident species has a slight competitive advantage (Table 4.6).

Example 4.5   Competition between two genotypes of *Arabidopsis thaliana* IV

Assume again that the two genotypes of *Arabidopsis thaliana* (*Nd-1* and *C24*) investigated in example 4.1 and 4.2 are the two only genotypes in the *A. thaliana* population and that the density of *A. thaliana* is not controlled by other plant species. Furthermore, assume that mortality is high and density-independent and that both genotypes have the same probability of reaching reproductive age.

One of the pathogens known to attack natural *A. thaliana* populations is the biotrophic oomycete *Peronospora parasitica* (Holub et al. 1994). The pathogen causes downy mildew, but the effects of the disease in natural plant communities are unknown. The pathogen *P. parasitica* grows on a wide range of crucifers (Dickinson and Greenhalgh 1977) and may be considered a generalist and it is here assumed that the local population size of the pathogen is constant on an ecological time scale. Genotype *Nd-1* is susceptible to *P. parasitica* isolate *Cala2*, and genotype *C24* is resistant to the isolate (Holub and Beynon 1997).

The competition experiment described in Example 4.1 was repeated in the greenhouse both in the absence of the pathogen and where each seedling was infected with about hundred *P. parasitica* (*Cala2*) conidia (Damgaard and Jensen 2002). The fecundity was estimated from

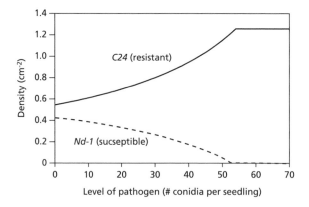

Fig. 4.5   Stable equilibrium densities (equilibrium 4.11d) of *A. thaliana* genotypes as a function of the pathogen level (h, measured in number of *P. parasitica* conidia per seedling) calculated using the maximum likelihood estimates of the competition model (Table 2) . The probabilities of reaching reproductive age were assumed to be 0.001.

dry weight data by linear regressions and fitted to competition model (4.9) with some a priori constraints on the parameter space since the disease had no effect on genotype *C24* (Damgaard 2003b). This experimental design is the minimum design required for applying the model. It would be beneficial to include more densities and proportions of the two species, and to include more levels of the pathogen (presence vs. absence data can hardly be described as a gradient). Nevertheless, the maximum likelihood estimates of parameters were used to predict the densities at equilibrium of the two *A. thaliana* genotypes as a function of the level of the pathogen (Fig. 4.5).

The statistical uncertainty of the estimated pathogen level when the susceptible genotype is just outcompeted at equilibrium was investigated using inequality (4.12) and Bayesian statistics (Fig. 4.6) and it is apparent that the degree of uncertainty of the pathogen level is high. A considerable part of the posterior distribution is outside the domain of the data (0-100 conidia per seedling) and the lack of certainty is probably due to the fact that there was only two pathogen treatments in the competition experiments. The results should therefore be interpreted with care.

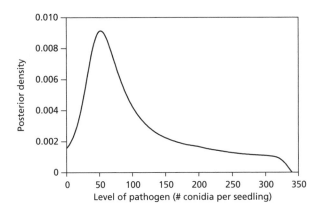

Fig. 4.6 Posterior distribution of the pathogen level (measured in number of *P. parasitica* conidia per seedling) when the susceptible *A. thaliana* genotype (*Nd-1*) is outcompeted by the resistant genotype (*C24*) assuming an uninformative prior distribution. The probabilities of reaching reproductive age were assumed to be 0.001.

Table 4.6 The invasion criteria of the four possible scenarios between an invading and a resident species with two possible scales of dispersing. Only three different strategies may invade if it is assumed that the resident species has a competitive advantage (Bolker and Pacala 1999).

		Resident species	
		Globally dispersing	Locally dispersing
Invading species	Globally dispersing	The probability that the invading species succeeds is adequately described by the mean-field model approximation.	A plant with a *colonisation strategy* may invade if the resident species has a clumped distribution.
	Locally dispersing	Both an *exploitation strategy*, where plants quickly exploits the limiting resource, and a *tolerance or phalanx strategy*, where plants are benefited in the inter-specific competitive interactions by an increase in the local density of conspecific plants, may invade.	The invading species may be present in local patches.

The three spatial strategies in Table 4.6, which may invade a habitat even if they are competitively inferior to a resident species, should not be confused with the CSR classification of plant strategies due to Grime (2001). In the CSR classification, it is assumed that the ecological success of different plant species in a habitat mainly is explained by the intensity of disturbance and the general productivity in a habitat (Table 4.7).

Table 4.7 The CSR classification of plant strategies (from Grime 2001). A plant species may have a variable amount of one of the three primary strategies which may de depicted in a De Finetti diagram (Grime's triangle).

	Productivity	
Intensity of disturbance	High	Low
Low	*Competitors*	*Stress-tolerators*
High	*Ruderals*	No viable strategy

In plant ecological research, the diversity of specific plant communities is often a central scientific question and predicting plant community structures by searching for possible community assembly rules since long has been a research goal in plant ecology. At least three research paths have been followed:

1) Either to test whether the species composition deviates from a specified null-model where species are assumed to be independent of each other (Conner and Simberloff 1979, Rees et al. 1996, Wilson et al. 1996).

2) To explain the species composition pattern by the ecology of different species groups (Weiher et al. 1998). If two plants species compete for the same limiting resources in a similar way, they are said to belong to the same functional type, which is a plant ecological term analogous to the guild or niche concept in the animal literature. The classification of plant species to functional types may be done by comparing either morphological and life history traits (e.g. Tilman 1997, Hooper 1998, Weiher et al. 1998), or the positions in the CSR classification or another strategy classification system (e.g. Westoby 1998). One hypothesis of a possible assembly rule of plant communities that has been tested is that the proportions between different functional types are constant (e.g. Wilson et al. 1996, Weiher et al. 1998, Symstad 2000).

3) Or analysing the stability properties of community matrices (Roxburgh and Wilson 2000b, a), which are matrices of the competition coefficients of n competing species:

$$C = \begin{matrix} 1 & c_{12} & c_{13} & \cdot & c_{1n} \\ c_{21} & 1 & c_{23} & \cdot & c_{2n} \\ \cdot & \cdot & \cdot & \cdot & \cdot \\ c_{n1} & c_{n2} & c_{n3} & \cdot & 1 \end{matrix} \qquad (4.14),$$

and similar approaches (e.g. Law and Morton 1996).

As discussed previously, good quantitative predictions of the future states of plant communities are highly in demand (e.g. Keddy 1990). However, it is discouraging that the ecological conclusions typically reached when the above methods are tried on actual cases are very general and unspecific on answering questions such as what will happen and when will it happen. There is still a lot of work to do!

5. Genetic ecology

Genetic variation

Until now mainly competitive plant – plant interactions during growth and development have been discussed, but conspecific plants in a limited geographic area also interact sexually by transferring pollen. The notion of a population is a convenient way of expressing our conviction that a group of plants share the same gene pool, either by sexual exchanges of gametes or by mixing the offspring in a geographic area through seed dispersal. Knowledge on the population structure is important because it is the population structure that determines how the genetic variation in a species is maintained and renewed, which again ultimately may influence the evolutionary fate of the species.

Contrary to the study of competitive plant growth, where a number of different growth models exist, the basic genetic model of inheritance has been well established for a hundred years. A sexual plant receives half of its chromosomes from the paternal gamete in the pollen and another complementary half-set from the maternal gamete in the egg cell. As a consequence of the well established basic model of inheritance, the notation and many of the calculations in population genetics has become standardised and are most practically made by using the free and user-friendly software that has been developed in recent years (e.g. http://www.biology.lsu.edu/general/software.html).

Opposite many ecological studies, it is costly to make a genetic analysis of the entire studied population. Hence in population genetic studies usually only a limited number of individuals are sampled; from these statistical inferences of the whole population are made. Three types of genetic variation are recorded and analysed using different models; either

1) The variation among alleles at a locus, e.g., isozymes, RFLP, RAPD etc., the allele variation is analysed using the infinite allele model, where each locus is assumed to have an infinite number of alleles (e.g. Nei 1987, Balding et al. 2001).

2) Variation in the number of small DNA repeats (microsatellites). Microsatellite variation is usually analysed using a stepwise mutation model (Valdes et al. 1993), or the infinite allele model.

3) Sequence variation, where the nucleotide sequence along a string of DNA of variable length is determined. The infinite site model, where the strings of DNA are assumed to be infinitely long (e.g. Nei 1987, Balding et al. 2001), is normally used to describe the variation between nucleotide sites. However, non-recombining DNA sequences (haplotypes) may be analysed using the infinite allele model.

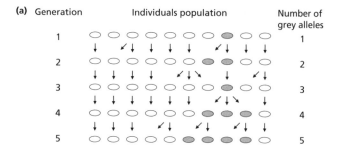

(a) Generation   Individuals population   Number of grey alleles

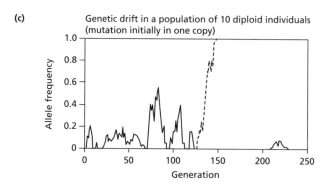

(b)

Genetic drift in a population of 100 diploid individuals (mutation initially in one copy)

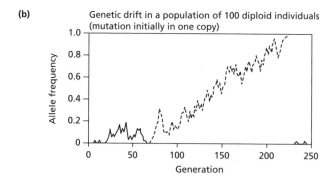

(c)

Genetic drift in a population of 10 diploid individuals (mutation initially in one copy)

Fig 5.1 Genetic drift. (a) The transmission of alleles from one generation to the next. A new allele that arises in generation 1 as a single mutation is present in four copies in generation 5, but the number fluctuates as a result of random differences in transmission every generation. (b, c) Examples of computer simulations of finite populations of two different sizes, showing the allele frequency fluctuations of new neutral alleles that arise. Note that most new alleles quickly go extinct soon after they arise (solid lines show repeated new alleles arising by mutation), but occasional an allele can increase by chance and become fixed in the population (dashed lines).

Figure after Silvertown and Charlesworth (2001).

The different descriptions of genetic variation complement each other and basically can produce the same conclusions, but the notation and calculation methods differ considerably. Usually the genetic variation used for making inferences on the population structure is assumed to be selectively neutral, which simplifies the calculations considerably.

Imagine a finite sexual plant population of N individuals at the reproductive age in ecological equilibrium; so that in each generation there is N individuals at the reproductive age, then each individual *on average* will contribute one paternal and one maternal successful gamete to the next generation. However, each plant will, depending on the species, produce a large number of paternal and maternal gametes that each has a small probability of being successful. If all plants produce the same number of gametes with the same probability of being successful, then the number of successful gametes per plant will be Poisson distributed, i.e., many plants will contribute zero, one or two gametes and fewer will contribute several gametes to the next generation. As a consequence of the variable number of successful gametes, selectively neutral alleles may either increase or decrease in frequency in the population due to random stochastic events (Fig. 5.1). This phenomenon that alleles may vary in frequency due to stochastic events during reproduction from one generation to the other is known as genetic drift (e.g. Wright 1969, Crow and Kimura 1970, Gale 1990), which is important in small and intermediate sized populations (Fig. 5.1).

The importance of genetic drift increases with the variance in number of successful gametes, inbreeding, and decreasing population size (Caballero and Hill 1992a). Since plants vary considerably in their capacity to produce gametes and the probability of being a successful gamete depends on the spatial setting of plants and other factors, the number of successful gametes per plant in a natural population will typically be more variable than a Poisson process. It has become a standard practice to measure the effect of genetic drift in a natural population by the effective population size. The effective population size is the size of an ideal (Fisher-Wright) population that would experience the same amount of genetic drift as the studied natural population (Wright 1969, Crow and Kimura 1970, Nei 1987). The effective population size, N_e, of a random mating population of constant census size, N, is:

$$N_e = \frac{4N}{2 + V_k^2}$$
(5.1),

where V_k^2 is the variance of the number of successful gametes (Caballero and Hill 1992a). In the case of a Poisson distributed number of successful gametes, $V_k^2 = 2$, the effective population size is equal to the census population size.

Fluctuating population sizes increase the amount of genetic drift compared to what would be expected by a constant population size at the arithmetic mean of the fluctuating population size (Crow and Kimura 1970). This effect may be especially critical for annual plant species, which are known to have quite fluctuating population sizes. However, the negative effect of fluctuating population sizes on the maintenance of neutral genetic variation is reversed by the presence of a seed bank with variable dominance (Nunney 2002).

Due to the rapidly increasing amount of DNA sequence variation data, it has become increasingly popular to interpret the genetic variation of a gene by its gene tree, which is constructed from a sample of DNA sequences, and the underlying coalescent process (Nordborg and Innan 2002). In the absence of recombination, a present sample of alleles will all be copies of the same allele that existed at some point backward in time. In a gene tree the process of coalescence of the sampled alleles can be followed backward in time until only one allele is left. Given an ecological model on how descendants are produced, it is possible to quantify the expected distribution of branch lengths in the gene tree and estimate model parameters from the observed distribution of branch lengths in the gene tree (Fig 3.9) (Beerli and Felsenstein 2001, Emerson et al. 2001, Nordborg and Innan 2002).

The amount of selectively neutral genetic variation in an isolated population is a balance between the production of new variation by mutation and the loss of variation due to genetic drift. Unfortunately, both the neutral mutation rate and the effective population size is generally unknown, but the product of the two may under certain equilibrium assumptions be determined from the observed number of different alleles, k, in a sample of n diploid individuals from the population:

$$E(k) = \frac{\theta}{\theta} + \frac{\theta}{\theta + 1} + \frac{\theta}{\theta + 2} + \cdots + \frac{\theta}{\theta + 2n - 1} \qquad (5.2),$$

where $\theta = 4N_e\mu$ is the product of the effective population size, N_e, and the neutral mutation rate, μ, (Ewens 1972). Furthermore, if k different selectively neutral alleles are found in the sample, the expected probability distribution of the number of alleles is given by

$$P(n_1, n_2, \cdots n_k \mid 2n, k) = \frac{2n!}{k! l_k n_1 n_2, \cdots n_k} \tag{5.3},$$

where l_k is a Sterling number of the first kind (Ewens 1972). Alternatively, θ may be determined by constructing the gene tree based on the DNA sequences of the sample (Kuhner et al. 1995).

If k different alleles with frequencies p_i are found in a sample of N diploid individuals from a population, the observed amount of genetic variation at the locus is usually quantified by the gene diversity, H, which is the unbiased probability that two randomly chosen alleles will be different:

$$H = \frac{2N}{2N-1} (1 - \sum_{i=1}^{k} p_i^2) \tag{5.4},$$

(Nei 1987). An analogue measure of the genetic variation in a sample of DNA sequences is the nucleotide diversity, i.e., average number of nucleotide differences per site between two sequences.

Inbreeding
Many plant species, especially among annual plants (Fig 5.2), are partially or almost completely self-fertilising (Schemske and Lande 1985, Cruden and Lyon 1989, Barrett et al. 1996) and such an inbreeding mating system influence the genetic structure in the population (e.g. Crow and Kimura 1970, Christiansen 1999).

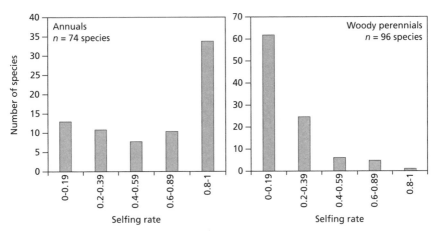

Fig. 5.2 The distribution of the probability of self-fertilising (selfing rate) in annual and woody perennial plant species. Figure after Barrett et al. (1996).

Consider a diploid unselected autosomal locus with only two alleles A and a in frequencies p and $q = 1 - p$, respectively, forming the genotypes AA, Aa and aa. In an ideal random mating population, where each female gamete is mated to a random male gamete (thus allowing for rare self-fertilisation events), the genotypes will be in Hardy-Weinberg proportions:

$$AA : \quad p^2$$
$$Aa : \quad 2pq \qquad (5.5).$$
$$aa : \quad q^2$$

In self-fertilising events homozygotes produce homozygotes but on average only half of the offspring of heterozygotes are heterozygotes, thus in a plant population with some degree of self-fertilising or another form of bi-parental inbreeding, the proportion of heterozygotes is reduced relative to Hardy-Weinberg proportions (5.5). The deficiency of heterozygotes relative to Hardy-Weinberg proportions is measured by the inbreeding coefficient, F:

$$AA : \quad p^2 + Fpq$$
$$Aa : \quad 2pq(1 - F) \qquad (5.6)$$
$$aa : \quad q^2 + Fpq$$

Inbreeding affects not only the genotypic distribution at a single locus, but also the way linked loci are associated in a population (Christiansen 1999). If the plant population has a constant selfing rate where a fraction α of new zygotes are produced by self-fertilisation and a fraction $1 - \alpha$ by random mating, the genotypic proportions will approach an equilibrium distribution with an equilibrium inbreeding coefficient of:

$$\hat{F} = \frac{\alpha}{2 - \alpha} \qquad (5.7),$$

(Haldane 1924). The selfing rate in plant populations may be inferred by determining the inbreeding coefficient using equation (5.7) under the assumption that the genotypic proportions are at equilibrium and the deviation from Hardy-Weinberg proportions only is due to self-fertilisation. Alternatively, the selfing rate may be determined more directly by making an offspring analysis of the allelic variation (e.g. Schoen and Lloyd 1992, Weir 1996).

The effective population size of a partially self-fertilising population is:

$$N_e = \frac{4N}{2(1 - F) + V_k^2(1 + F)}$$ (5.8),

where F is the inbreeding coefficient and V_k^2 is the variance of the number of successful gametes (Caballero and Hill 1992a, Wang 1997). In the case of Poisson distributed number of successful gametes, $V_k^2 = 2$, the effective population size is $N_e = N/(1 + F)$.

Population structure

A population can be defined as a group of plants that share the same gene pool, either by sexual exchanges of gametes or by mixing the offspring in a geographic area through seed dispersal. However, since the adult plants are sedentary and both pollen and seed dispersal often is skewed towards relatively small dispersal distances, e.g. dispersal models (3.11) and (3.12), it is clear that neighbouring plants, every thing else being equal, have a higher probability of sharing genes than more distantly positioned plants. Furthermore, in a continuous natural plant population, it is generally impossible to set the limits of when one population ends and the next population begins. The physical movement of seeds and pollen may be followed directly (Kjellsson et al. 1997) and the effect of dispersal distances on the genetic variance may be predicted (Wright 1969, Crawford 1984, Eguiarte et al. 1993). However, often it is more informative to use the distribution of genetic variation among the individuals in a population as a record of the way the individuals have interacted in the past.

In order to describe the population structure by a genetic analysis, it is necessary to notice an important sampling effect when a sample is made from a pooled subdivided population. If two random mating populations with different allele frequencies are pooled and erroneously regarded as one population, there will be a deficiency of heterozygotes relative to the Hardy-Weinberg proportions (Table 5.1). This sampling effect is called the Wahlund effect and becomes increasingly important with variation in allele frequencies among populations.

Table 5.1   The Wahlund effect in a diploid locus with two alleles. The genotypic frequencies in two different random mating populations and when the populations are pooled. The deficiency of heterozygotes due to the Wahlund effect may be described using (5.5) and since $(p_1{}^2 + p_2{}^2)/2 = \bar{p}^2 + F\,\bar{p}\,\bar{q} \Leftrightarrow F = (p_1 - p_2)2/(p_1 + p_2)(q_1 + q_2) \Leftrightarrow F = Var(p)/\bar{p}\,\bar{q}$.

	AA	Aa	aa
Population 1	$p_1{}^2$	$2\,p_1\,q_1$	$q_1{}^2$
Population 2	$p_2{}^2$	$2\,p_2\,q_2$	$q_2{}^2$
Population 1+2	$(p_1{}^2 + p_2{}^2)/2$ or $\bar{p}^2 + F\,\bar{p}\,\bar{q}$	$p_1\,q_1 + p_2\,q_2$ or $2\,\bar{p}\,\bar{q}\,(1 - F)$	$(q_1{}^2 + q_2{}^2)/2$ or $\bar{q}^2 + F\,\bar{p}\,\bar{q}$

Traditionally the population structure is investigated in a design where the genetic variation in a geographical area is sampled from a number of small areas, where each area is so small that the individuals in the local area (the subpopulation) may be assumed to share a common gene pool. The population structure in the overall or total sample is analysed using Wrights F-statistic where the genetic variation is partitioned within and among subpopulations (Wright 1969, Nei 1987, Wang 1997, Nagylaki 1998, Balding et al. 2001). In the total sample there may be a deficiency of heterozygotes relative to Hardy-Weinberg proportions, F_{it}, due to the combined effect of possible inbreeding within the subpopulations, F_{is}, and a possible Wahlund effect, F_{st}. The deficiency of heterozygotes in the total sample is related to subpopulation inbreeding and the Wahlund effect by:

$$1 - F_{it} = (1 - F_{is})(1 - F_{st})$$
(5.9),

(e.g. Nei 1987).

If the subpopulations are distributed as a mosaic structure connected by relatively rare migration events of gametes or seeds among subpopulations, it is common to refer to the collection of subpopulations as a metapopulation (Hanski 1998). Given a model of migration among the subpopulations, e.g. an island model or a stepping stone model, including the probability of extinction and re-colonisation of subpopulations the migration rate may be inferred from the estimated value of F_{st} (Pannel and Charlesworth 2000, Balding et al. 2001). Alternatively, the migration pattern of the metapopulation may be inferred from the gene tree of the sampled genetic variation (Beerli and Felsenstein 2001, Nordborg and Innan 2002).

The amount of genetic drift, and consequently the amount of selectively neutral genetic variation, depends highly on the population structure, but the usual effect of population subdivision is to increase the importance of genetic drift (Whitlock and Barton 1997). Migration prevents the build up of genetic differences between populations and even a small amount of migration is sufficient to prevent appreciable genetic differences among subpopulations (Nei 1987). However, if the migration events are very rare, genetic differences will gradually build up and it may be useful to determine the phylogeny from DNA sequences in order to test various hypotheses on when and in what order the populations separated (Nei 1987).

Example 5.1   Population structure of *Melocactus curvispinus*

The genetic variation in 17 polymorphic isozyme loci in the partially self-fertilising cactus *Melocactus curvispinus* was measured in a hierarchical geographic design in Venezuela (Nassar et al. 2001). The cactus has an estimated mean selfing rate of 0.24 (estimates from individual plants range between 0 and 0.82), which is expected to give an equilibrium inbreeding coefficient of $0.24/(2-0.24) = 0.136$ (see equation (5.7)). The dispersal distances of pollen and seeds are relatively short. Cacti were sampled at nine regions across Northern Venezuela and in two of the nine regions the regional genetic variation was investigated at five local areas (Table 5.2). There was a significant effect of inbreeding (F_{is}), but some of the deficiency in heterozygotes within subpopulations may be due to the Wahlund effect within the designated subpopulations (Nassar et al. 2001). There was also a significant genetic differentiation among subpopulations (F_{st}) both at the regional level and across Northern Venezuela. This genetic differentiation among subpopulations is possibly due to genetic drift (Nassar et al. 2001).

Table 5.2   The mean and (standard error) of F_{is} and F_{st} of 17 polymorphic isozyme loci in the partially self-fertilising cactus *Melocactus curvispinus* (after Nassar et al. 2001)

Geographical range	F_{is}	F_{st}
Northern Venezuela	0.348 (0.077)	0.193 (0.047)
Region I	0.402 (0.087)	0.187 (0.059)
Region II	0.194 (0.074)	0.084 (0.036)

6. Natural selection

Mode of selection

While most of the genetic variation found in plant populations probably is selectively neutral (Kimura 1983), some genetic variation makes a difference. Some genotypes will code for a phenotype that in a given environment *on average*, due to e.g. faster growth or a better defence against parasites, will leave more offspring in the next generation.

Natural selection is composed of three principles (Lewontin 1970):

1) Different individuals in a population have different morphologies and physiologies (phenotypic variation).

2) Different phenotypes have different rates of survival and reproduction in different environments (differential fitness).

3) There is a correlation between parents and offspring in the contribution of each to future generations (fitness is heritable).

The fitness of a phenotype depends on many different characters that typically are correlated and determined by a combination of genotypic and environmental factors. The genotypic difference that causes the average increase in the number of descendants may be a single allele (a Mendelian character) or a number of alleles in a favourable combination (a quantitative character), and such alleles will on average increase in frequency due to natural selection (Wright 1969). The consequent increase in mean fitness caused by natural selection is equal to the genetic variance in fitness (Fisher 1958, Frank 1997).

Since the biological world has so many different life forms and reproductive strategies it is not operational to give a precise and complete definition of fitness (Lewontin 1970). Instead, it is convenient to define fitness relative to the studied organism. Where the ecological success of a plant species at a specific place only depended on fecundity and viability (chapters 2-4), the fitness of an individual sexual plant depends on the number of gametes transmitted to the next generation by both the maternal and the paternal line. Thus the fitness of an individual sexual plant depends, besides viability and fecundity, on sexual – and gametic selection (Fig. 6.1). The selective forces on a plant species without sexual transmission, i.e. apomictic or purely self-fertilising plants, operate on the whole genome and the evolutionary process may be modelled using ecological models as exemplified above in the case of competitive interactions between different *Arabidopsis thaliana* genotypes.

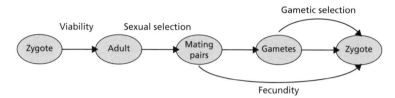

Figure 6.1   A generalised life cycle showing the components of selection.
Figure after Hartl and Clark (1989)

Male sexual selection may occur when the diploid male gamete donor can manipulate the probability of donating a male gamete to a successful zygote. For example, it may be a selective advantage to increase the amount of pollen produced in a wind-pollinated species, or to increase the nectar reward in an insect-pollinated species. Foreign pollinated plants have little control on the genetic makeup of the pollen arriving at the stigma, but there may be a possibility for female sexual selection and male gametic selection among male gametophytes growing down the style (Stephenson and Bertin 1983, Queller 1987, Walsh and Charlesworth 1992). When the pollen settles on the stigma, a male gametophyte "germinates" and grows inside the style until it reaches an ovary and fertilises an ovule. In the cases where more pollen lands at the same time on the stigma there may be a race among the male gametophytes to reach an ovule. A surprisingly large amount of the total genome seems to be expressed in the male gametophyte; about 60% of the structural genes expressed in the sporophyte are also expressed in the male gametophyte and exposed to selection (Hormaza and Herrero 1992, Mulcahy et al. 1996). In addition, 5 – 10 % of the genes in the genome is only expressed in the male gametophyte (Hormaza and Herrero 1992). The effect of gametic selection may potentially be important on rare recessive deleterious alleles, because they cannot be "hidden" in heterozygotes (Haldane 1932), although the importance decreases with the selfing rate (Damgaard et al. 1994). In the development of the seed there is a possibility that the diploid maternal plants selectively aborts ovules or developing seeds and fruits, as a combination of female gametic selection and early viability selection (Stephenson and Bertin 1983).

When measuring total or lifetime fitness of different phenotypes, it is tempting to census the population at e.g. the reproductive age and determine the number of each phenotype. However, if selection is operating both before and after the time of censusing, then this procedure may give a biased estimate of the phenotypic fitness (Prout 1965).

An allele may affect the fitness components in different ways. An allele may have a positive effect on plant viability and fecundity, but has no effect on the male sexual selection (e.g., attractiveness to pollinating insects). Conversely, a colour mutation that make the flower more attractive to pollinating insect may have a positive effect on the male sexual selection but no effect on viability. Perhaps there may even be a trade-off (or conflict Partridge and Hurst 1998), so that a mutation that increases attractiveness to pollinating insects and thereby increases male sexual selection also increases the probability that herbivorous insects visit the plant so that viability or fecundity is reduced.

Possibly as an adaptation to the sedentary adult life form, many plant species are plastic and are able to respond to the environment. For example a plant that grows in the shade of forest trees will typically grow taller with reduced branching and produce thinner leaves than a conspecific plant grown under full light conditions. Different plant species may respond differently to the environment, i.e. have different norms of reactions. A specific type of reaction norm is to a certain extent inherited by the offspring and thus controlled to some degree by the genotype of the plant (e.g. Pigliucci et al. 1999). Since different genotypes may code for different reaction norms that may meet the challenges of the environment differently, natural selection may operate on the genes controlling the reaction norm.

Natural selection on a single locus

Since many plant species are partially or almost completely self-fertilising (Schemske and Lande 1985, Cruden and Lyon 1989, Barrett et al. 1996) and the evolutionary processes have been shown to depend critically on the selfing rate and the consequently reduced proportion of heterozygotes it is often imperative to include the selfing rate in the modelling of natural selection in hermaphroditic plant populations (e.g. Bennett and Binet 1956, Allard et al. 1968, Ohta and Cockerham 1974, Gregorius 1982, Caballero et al. 1992, Caballero and Hill 1992b, Charlesworth 1992, Pollak and Sabran 1992, Damgaard et al. 1994, Christiansen et al. 1995, Overath and Asmussen 1998, Christiansen 1999, Damgaard 2000, Rocheleau and Lessard 2000, Damgaard 2003a).

Consider a single autosomal locus with two alleles, A and a, in a partially self-fertilising hermaphroditic plant population with non-overlapping generations. The population has a selfing rate of α. Each generation starts by the production of zygotes from gametes (Fig 5.1). Before selection, the frequency of the A allele among male and female gametes is p_m and p_f respectively, and the frequency of the a allele among male and female gametes is $q_m = 1 - p_m$ and $q_f = 1 - p_f$, respectively. Assuming only weak selection, the coefficient of inbreeding is close to equilibrium (equation 5.6), and as the equilibrium proportions due to inbreeding are reached relatively quickly (Wright 1969, Nordborg and Donnelly 1997), then the proportions of the three genotypes before selection can be expressed as:

AA: $\quad x = (1 - \alpha)p_m\, p_f + \alpha\left(x_f + \dfrac{y_f}{4}\right)$

Aa: $\quad y = (1 - \alpha)(p_m\, q_f + q_m\, p_f) + \alpha\left(\dfrac{y_f}{2}\right)$ $\qquad$ (6.1),

aa: $\quad z = (1 - \alpha)\, q_m\, p_f + \alpha\left(z_f + \dfrac{y_f}{4}\right)$

where $x_f = p_f^2 + \hat{F}\, p_f\, q_f$, $y_f = 2p_f\, q_f - 2\hat{F}\, p_f\, q_f$ and $z_f = q_f^2 + \hat{F}\, p_f\, q_f$ (e.g. Damgaard 2000). When selection is strong the assumption that the genotypic proportions are in equilibrium with respect to the selfing rate does not hold and it is necessary to model the change in the inbreeding coefficient as a response to selection among genotypes (Overath and Asmussen 1998, Rocheleau and Lessard 2000).

The frequency of the A allele in the population before selection is:

$$p = x + \frac{y}{2} = \frac{p_f + \alpha\, p_f + (1 - \alpha)p_m}{2} \qquad (6.2).$$

For simplicity, only genotypic selection is assumed to operate in the system, and the relative fitness of the three genotypes is shown in table 6.1. Viability selection is assumed to take place before reproduction. Fecundity selection operates on the number of seeds produced by the plants. Male sexual selection operates on the probability of donating a male gamete to a successful outcrossing event, disregarding the pollen needed for self-fertilising events. Note that the term fecundity selection normally includes gametic selection during zygote formation (Fig. 6.1), but since gametic selection is assumed not to take place, the selective forces can be adequately described by the frequencies of the alleles in the female and male gamete pool.

Table 6.1 The relative fitness of the three genotypes.

Genotype	Viability selection	Fecundity selection	Male sexual selection
AA	$1 + s_v$	$1 + s_f$	$1 + s_m$
Aa	$1 + h_v\, s_v$	$1 + h_f\, s_f$	$1 + h_m\, s_m$
aa	1	1	1

After selection, the frequency of A among the male gametes is determined by the combined effect of viability selection and male sexual selection, which are assumed to be multiplicative (Penrose 1949, Bodmer 1965, Feldman et al. 1983, Damgaard 2000):

$$p'_m = \frac{x'_m + \frac{1}{2} y'_m}{x'_m + y'_m + z'_m} \tag{6.3a}$$

where, $x'_m = x(1 + s_v)(1 + s_m)$, $y'_m = y(1 + h_v\, s_v)(1 + h_m\, s_m)$ and $z'_m = z$. Likewise, after selection, the frequency of A among the female gametes is determined by the combined effect of viability selection and fecundity selection, which again are assumed to be multiplicative:

$$p'_f = \frac{x'_f + \frac{1}{2} y'_f}{x'_f + y'_f + z'_f} \tag{6.3b}$$

where, $x'_f = x(1 + s_v)(1 + s_f)$, $y'_f = y(1 + h_v\, s_v)(1 + h_f\, s_f)$ and $z'_f = z$. Note that, viability selection on hermaphroditic plants affects both sexes equally, whereas fecundity selection and male sexual selection affects the two sexes differently. The expected frequency of the A allele among the newly formed zygotes in the next generation, p', can be calculated by inserting p'_m and p'_f into equation (6.2).

In the simple case of an outcrossing population with viability selection and sex symmetric selection, i.e. fecundity and male sexual selection are equal, the model is identical to the standard viability selection model (Bodmer 1965). The expected change in allele frequency is then:

$$p' - p = pq \, \frac{(w_{AA} - w_{Aa})p + (w_{Aa} - w_{aa})q}{\bar{w}} \tag{6.4},$$

where w_{ij} is the fitness of genotype ij and $\bar{w}$ is the mean fitness in the population. In an effectively infinite population with no genetic drift there are three stable equilibria of recursive equation (6.4); either allele A or a becomes fixed, or in the case of overdominance, $w_{AA} < w_{Aa} > w_{aa}$, there is a stable polymorphism (e.g. Hartl and Clark 1989).

The assumption of sex symmetric selection will in many cases be biologically unrealistic, since the fitness components of the different genders in hermaphroditic plants are only weakly correlated. Although it is difficult to measure paternal fitness, it seems that fecundity and paternal fitness is not strongly correlated and in some cases may be negatively correlated (Campbell 1989, Ross 1990). Plant viability and fecundity strongly depend on abiotic and biotic factors, whereas the variation in paternal function, at least in some cases, is observed to be lower than the variation in female functioning (Stephenson and Bertin 1983, Mazer 1987, Schlichting and Devlin 1989). Additionally, selective forces may change the allocation of resources between the male and female parts of hermaphroditic plants (Lloyd and Bawa 1984). For example, in gynodioecious plant populations the sex allocation in hermaphrodites has been shown to depend on the frequency of females and the mode of inheritance (e.g. Gouyon and Couvet 1987).

The case of sex asymmetric selection in an effectively infinite population has been studied in detail for outcrossing populations (e.g. Bodmer 1965, Feldman et al. 1983, Selgrade and Ziehe 1987, Christiansen 1999). Generally, the relaxation of assumption of sex symmetry may give qualitatively new results with multiple stable equilibria in the cases where the selective forces are opposite in the two sexes (Bodmer 1965). The combined effect of sex asymmetric selection and self-fertilisation has not been explored in detail. However, from a limited numerical study of the system, it is obvious that even in very simple genetic systems the combined effect of sex asymmetric selection and self-fertilisation is important for the rate of evolution (Fig. 6.2) (Damgaard 2000).

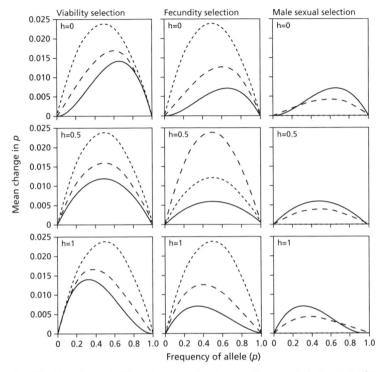

Fig. 6.2 **The effect of different modes of selection. The expected change in frequency,** $M = p' - p$, **of an advantageous allele as a function of the allele frequency at different selfing rates (full line:** $\alpha = 0$, **dashed line:** $\alpha = 0.5$, **dotted line:** $\alpha = 1$) **and dominance levels** (h). **Viability selection:** $s_v = 0.1$, $s_f = 0$, $s_m = 0$. **Fecundity selection:** $s_v = 0$, $s_f = 0.1$, $s_m = 0$. **Male sexual selection:** $s_v = 0$, $s_f = 0$, $s_m = 0.1$.
Figure after Damgaard (2000).

In the above model it is implicitly assumed that the selective forces operate on a single locus under consideration independent of the other loci. However, natural selection operates on more loci, which are more or less tightly associated (e.g. Barton and Turelli 1991, Christiansen 1999, Bürger 2000). The association of alleles at different loci is a balance between selection or genetic drift, which will cause alleles to be statistically associated, and recombination which will break the established associations (Box 6.1). This genetic association makes the exact modelling of the natural selection process on allele frequencies complicated, especially in hermaphroditic plants with an intermediary selfing rate. An important phenomenon that is caused by the physical linkage of loci is that neutral or even deleterious genes may increase in frequency if they are associated to a selected advantageous allele (hitchhiking selection) (e.g. Christiansen 1999).

91

Box 6.1 The association of linked loci

Consider two physically linked loci with two alleles, *A*, *a* and *B*, *b*. If genotypes that have both the *A* and the *B* allele are positively selected, the frequency of *AB* gametes after selection will be higher than expected if the alleles were statistically independent, i.e. at Robbins proportions: $P(AB) = P(A)P(B)$. Likewise if genotypes that have both the *A* and the *B* allele randomly produced more offspring (genetic drift), the frequency of *AB* gametes after reproduction would be higher than expected. The deviation from the expected Robbins proportions may be quantified by the linkage disequilibrium, which in the two-loci case is: $D = P(AB) - P(A)P(B)$.

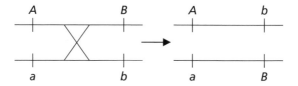

Recombination *on average* will tend to break up associations between alleles and the rate of convergence to the Robbins proportions will increase with the rate of recombination. However, the rate of convergence is a decreasing function of the selfing rate and in a completely self-fertilising plant species recombination has no effect on the statistical association of alleles in the population (Christiansen 1999, Balding et al. 2001).

Finite populations

Often the effect of genetic drift cannot be excluded either because the population is not effectively infinite or the allele is rare, e.g. after a mutation or a migration event. In such cases there are only two equilibrium states of an allele, either it becomes fixed or it is lost. Evolution in finite populations can be viewed as the ongoing fixation of new advantageous alleles and the rate of evolution will be controlled by selection and genetic drift acting on new advantageous alleles (Otto and Whitlock 1997). The fate of an advantageous allele in a finite population depends on the selective advantage of the allele, its coefficient of dominance, the effective population size, the demography of the population, the mode of selection and the degree of inbreeding (Haldane 1927, Caballero and Hill 1992b, Charlesworth 1992, Pollak and Sabran 1992, Otto and Whitlock 1997, Damgaard 2000).

Consider again the weakly selected autosomal locus with two alleles, A and a discussed in the previous section, but now in a plant population of census size N. The expected fixation probability and the mean fixation time can be calculated from the expected change in allele frequency, $M = p' - p$, and the mean variance of the change, $V = p(1 - p)/2N_e$, using the diffusion approximation (Kimura 1962).

If we assume that initially only a single copy of the A allele is present, the fixation probability of the A allele is calculated by:

$$P = \frac{\int_0^{1/2N} G(x)\, dx}{\int_0^1 G(x)\, dx} \tag{6.5}$$

where $G(x) = \exp\left(-\int (2M/V)\, dx\right)$ (Kimura 1962) (Fig 6.3).

Correspondingly, the mean time to fixation is calculated by:

$$T = \int_{1/2N}^1 \frac{4N_e P}{x(1-x)\, G(x)} \left(\int_x^1 G(t)\, dt\right) dx \tag{6.6}$$

(Ewens 1963, 1969).

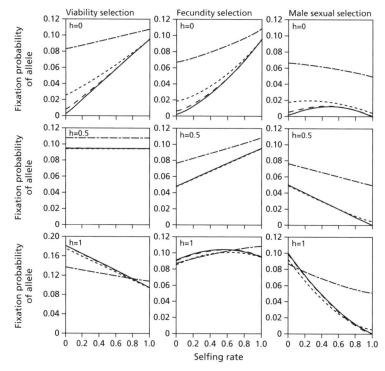

Fig. 6.3 The expected fixation probability of an advantageous allele (see Fig. 6.2) as a function of the selfing rate for different census population sizes (full line: N = 10,000, dashed line: N = 1000, dotted line: N = 100, dash-dotted line: N = 10) and dominance levels (h). Viability selection: s_v = 0.1, s_f = 0, s_m = 0. Fecundity selection: s_v = 0, s_f = 0.1, s_m = 0. Male sexual selection: s_v = 0, s_f = 0, s_m = 0.1. Figure after Damgaard (2000).

Density dependent selection

It is a common simplification to separate ecological and evolutionary processes by assuming that ecological processes occur faster than evolutionary processes. However, sometimes the ecological and evolutionary processes depend on each other, so that neither ecological nor evolutionary processes can be understood without considering both processes at the same time (Heino et al. 1998). A classic example is the coevolution of a parasite and its host (May and Anderson 1983). Generally, when selection depends on the frequencies or densities of genotypes, evolutionary processes depend on ecological processes (Cockerham et al. 1972, Slatkin 1979, Asmussen 1983b, Asmussen and Basnayake 1990, Christiansen and Loeschcke 1990). Neglecting the importance of ecological effects may lead to erroneous conclusions regarding the evolutionary forces. For example, a case of intraspecific competition in a density-dependent setting may be interpreted as frequency-dependent selection in a standard population genetic model, because such models implicitly assume that the population is growing exponentially.

An advantageous allele may have a positive effect on one or more of the standard fitness components (Fig 6.1). However, it may also affect the number of individuals that can be sustained in a certain environment at ecological equilibrium (carrying capacity) or the intraspecific competitive ability of a genotype (Joshi 1997). For example, consider a plant population that is attacked by a pathogen, in which an allele that confers resistance towards the pathogen is introduced. Resistant plants are not attacked by the pathogen and consequently grow faster than susceptible plants. Faster growth may lead to larger plants that produce more seeds (fecundity selection) and consequently have a higher intrinsic population growth rate at low densities. Faster growth may also give the resistant plants an intraspecific competitive advantage at intermediary and high density and possibly also an improvement in interspecific competitive ability, which consequently may increase the carrying capacity of the resistant plants.

If the genetic system is simple, i.e. the phenotypic effects of the alleles are additive and codominant and the genotypes are in Hardy-Weinberg proportions, the evolutionary forces may be analysed in a Lotka-Volterra type model of species interactions, e.g. recursive equation (4.2), (Hofbauer and Sigmund 1998). Alternatively, the expected frequency change due to natural selection of alleles that alter the intraspecific competitive ability and/or the carrying capacity may be analysed in a density-dependent selection model. There exist various models that incorporate some ecological realism into standard population genetic selection models (reviewed by Christiansen 2002). One of these is the hyperbolic density-dependent selection model with intra-specific competition (Asmussen 1983b), which is suitable for plant populations because 1) It allows a genotypic description of selection that is readily generalised to include partial inbreeding using F-statistics, 2) It takes both intra-specific competition and density-dependent selection into account, both of which have been shown to play an important role in evolutionary processes (e.g. Winn and Miller 1995, Joshi 1997, Smithson and Magnar 1997), 3) The hyperbolic density-dependent fitness model has relatively simple equilibrium properties (Asmussen 1983b) and has been shown to fit plant data well as discussed in chapter two.

Consider again the single autosomal locus with two alleles, A and a, in a partially self-fertilising hermaphroditic plant population of N individuals at reproductive age (adult census) with non-overlapping generations. The population has a selfing rate of α. Each generation starts by the production of zygotes from gametes (Fig 5.1). Before selection, which for simplicity is assumed to be sex symmetric, the frequency of the A allele is p, and the frequency of the a allele is $q = 1 - p$. Assuming only weak selection, the population has an equilibrium inbreeding coefficient (equation 5.6), and since the equilibrium proportions due to inbreeding are reached relatively quickly (Wright 1969, Nordborg and Donnelly 1997), then the proportions of the three genotypes before selection and reproduction can be expressed as:

$$AA : \quad x = p^2 + \hat{F} pq$$

$$Aa : \quad y = 2pq(1 - \hat{F}) \tag{6.7},$$

$$aa : \quad z = q^2 + \hat{F} pq$$

The individuals mate and produce offspring. The probability that an offspring survives to reproductive age the following year depends on its genotype. The absolute fitnesses, w_{ij}, of the three genotypes are assumed to depend on the size of the population and the frequencies of the other genotypes according to a hyperbolic density-dependent selection model with intra-specific competition (Asmussen 1983b):

$$w_{AA} = \frac{1 + r_{AA}}{1 + r_{AA} \, (x + c_{AA,Aa} \, y + c_{AA,aa} \, z) \, N/K_{AA}}$$

$$w_{Aa} = \frac{1 + r_{Aa}}{1 + r_{Aa} \, (c_{Aa,AA} \, x + y + c_{Aa,aa} \, z) \, N/K_{Aa}} \tag{6.8},$$

$$w_{aa} = \frac{1 + r_{aa}}{1 + r_{aa} \, (c_{aa,AA} \, x + c_{aa,Aa} \, y + z) \, N/K_{aa}}$$

where r_{ij} is the intrinsic rate of population growth of genotype ij, K_{ij} is the carrying capacity (or number of occupied sites at ecological equilibrium Lomnicki 2001) of genotype ij, and $c_{ij,kl}$ is the competition coefficient of genotype kl on genotype ij. The competitive interactions are modelled using the mean-field approach (Levin and Pacala 1997). When all $c_{ij,kl}$ are set to one, all genotypes have equal competitive ability, thus the model is reduced to a density-dependent selection model. When N is set to zero, the model reduces to a standard viability selection model.

The average of the absolute fitness value weighted by the genotypic proportions equals the population growth rate (Asmussen 1983b). Consequently, in the next generation, the expected number of individuals at reproductive age and the frequency of the A allele in the population will be:

$$N' = N(x \, w_{AA} + y \, w_{Aa} + z \, w_{aa})$$ (6.9a)

$$p' = \frac{x \, w_{AA} + \frac{1}{2} y \, w_{Aa}}{x \, w_{AA} + y \, w_{Aa} + z \, w_{aa}}$$ (6.9b)

The random mating case is analysed by Asmussen (1983b), and the purely self-fertilising case can be compared to the haploid case analysed by Asmussen (1983a). Generally, inbreeding has an effect on the evolution of alleles affecting population ecological characteristics. For example, in an underdominant case the number of stable internal equilibria decreased from two to one with only a slight degree of inbreeding. Equilibrium frequencies of stable internal equilibria and stability of fixation equilibria may also be effected by the degree of inbreeding (Damgaard 2003a).

Measuring natural selection in natural populations

Previously, the expected changes in allele frequencies due to natural selection has been described assuming the force of selection operating on the locus is known. However, this will rarely be the case; instead the selective forces need to be inferred by either:

1) Monitoring the changes in allele frequencies caused by natural selection (e.g. Manly 1985).
2) Measuring different phenotypic components of fitness and establishing a relationship between the fitness of different phenotypes and the expected evolutionary change in allele frequencies (Lande and Arnold 1983).
3) Testing deviations in allele frequencies or DNA-sequence patterns from a null model of selective neutrality. A rejection of the null hypothesis of selectively neutrality will indicate that selection has been operating in the past, but not necessarily the mode or the magnitude of selection (e.g. Golding 1994, Nielsen 2001, Nordborg and Innan 2002).

Monitoring allele frequency changes

In principle, monitoring the changes in allele frequencies would give a good estimate of the selective forces operating on the locus (Bundgaard and Christiansen 1972, Manly 1985). The problem is that different loci are linked and it is difficult to assess the selective forces to a particular locus. However, in experimental populations it is possible to examine the selective forces operating on different alleles in the same genetic background either by a crossing scheme or by genetic manipulation.

Phenotypic fitness

Natural selection operates on the phenotypes, which are comprised of multiple characters, and the joint distribution of phenotypic values is changed *within* a generation regardless of their genetic basis (Lande and Arnold 1983). The change in the mean values of multiple measured phenotypic characters produced by selection within a generation (the selection differential) may be described by the covariance between relative fitness and phenotypic values

$$s = \bar{z}^* - \bar{z} = \mathrm{Cov}(w(z), z) \tag{6.10},$$

where z is the vector of phenotypic values before selection and $w(z)$ is the vector of the corresponding relative fitnesses; $\bar{z}$, $\bar{z}^*$ are the mean phenotypic values before and after selection, respectively (Lande and Arnold 1983). Next, realising that multiple characters almost always are correlated genetically or phenotypically the change in the mean phenotypic values *across* one generation may be expressed by:

$$\Delta z = GP^{-1}s \tag{6.11},$$

where G,P are the covariance matrices of the additive genetic effects and the phenotypic values (the sum of additive genetic effects and independent environmental effects including non-additive genetic effects due to dominance and epistasis), respectively (Lande and Arnold 1983). The vector of the partial regression coefficients of the relative fitness of each character is

$$\beta = P^{-1}s \tag{6.12},$$

which may now be interpreted as the influence of each character on the relative fitness after the residual effect of the other characters have been removed (Lande and Arnold 1983). That is, from a covariance matrices of the phenotypic values and individual fitness measurements, the selective forces operating on the character may be estimated (example 6.1).

Example 6.1   Natural selection in *Impatiens capensis*

Five characters (seed weight, germination date, plant size in June, early growth rate, and late growth rate) and fitness (final adult size) were measured in a natural population of the annual plant *Impatiens capensis* (Mitchell-Olds and Bergelson 1990). Since the six measurements were made sequentially in time a causal path analysis could be made, e.g. seed weight may affect the size of the plant in June, but the size of the plant in June cannot have any effect on the weight of the seed from which the plant germinated (Fig 6.4).

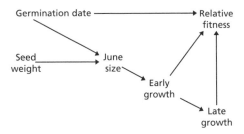

Fig 6.4 Path diagram of statistically significant predictor variables affecting fitness in Impatiens capensis.

Figure after Mitchell-Olds and Bergelson (1990).

The path analysis revealed both how much a character affected fitness directly (calculated using equation (6.12)) and indirectly by affecting other characters that affected fitness (Table 6.2). For example, the size of the plant in June had almost no direct effect on relative fitness, but a large indirect effect on relative fitness because the size of the plant in June had an effect on early growth.

Table 6.2   The relationship between characters and relative fitness in *Impatiens capensis*. All estimates are standardised. The indirect effects on relative fitness are calculated from all causal paths from a character via each "subsequent" character (Fig. 6.4). Total causal effects on relative fitness are the sum of the direct effects and the indirect effects (Mitchell-Olds and Bergelson 1990).

Character	Selection differential	Direct causal	Indirect causal	Total causal
Seed weight	0.13	0.04	0.10	0.14
Germination date	-0.36	-0.08	-0.32	-0.40
June size	0.53	-0.02	0.52	0.50
Early growth rate	0.73	0.37	0.38	0.76
Late growth rate	0.74	0.48	–	0.48

The fitness regression method of Lande and Arnold (1983) is a convenient way of measuring natural selection, but the method hinges on a number of important assumptions, e.g.:

1) That the fitness relevant characters have been measured. If an important fitness relevant character has not been measured but is correlated to an unselected measured character, then erroneous conclusions on the selection operating on the measured character will be made (Lande and Arnold 1983, Mitchell-Olds and Shaw 1987).
2) That there are no genotype-environment interactions (Lande and Arnold 1983). This assumption is probably particular critically for plant populations due to the adult sedentary life form. For example, significant genotype-environment interactions for fitness have been observed down to a spatial scale of a few meters in *Erigeron annuus* (Stratton 1995).
3) The shape of the fitness surface, e.g. multiple modes, stabilising selection, and balancing selection, is not always appropriately described by the regression coefficients (Mitchell-Olds and Shaw 1987, Schluter and Nychka 1994), but see Schluter and Nychka (1994) for a more general curve fitting approach.

Never the less, the fitness regression method may be used as an effective tool to generate testable hypotheses about the forces of natural selection operating on a character (Mitchell-Olds and Shaw 1987).

Deviation from neutrality

The case of selective neutrality is a theoretically relative simple case and a number of theoretical expectations on the distribution of genetic variation have been established. Therefore it has been attractive to demonstrate the effect of natural selection by comparing the observed genetic variation with what would be expected under the null-hypothesis of selective neutrality (e.g. Nielsen 2001).

For example, the hypothesis that alleles are maintained in the population due to heterozygote advantage may be tested by comparing the observed gene diversity (5.4) with the expected frequency distribution of neutral alleles (5.3) (Watterson 1977). Another important example is the difference between substitution rates of DNA-nucleotides that either have no effect on the resulting protein (synonymous substitutions), or change the amino-acid composition of the resulting protein (non-synonymous substitutions) (McDonald and Kreitman 1991, Nielsen 2001) (example 6.2). Generally, natural selection will affect the process of coalescence and the resulting gene tree, thus selection may be inferred by comparing observed gene trees with expected gene trees under neutrality. Although, it will often be necessary to sample the sequence variation in several additional unlinked loci in order to distinguish between demographic and selective processes (Nordborg and Innan 2002).

Example 6.2   Natural selection in *Arabidopsis thaliana*

The *PgiC* locus in *Arabidopsis thaliana*, which plays a role in the sugar metabolism, was sequenced (Kawabe et al. 2000). Among other statistics of the genetic variation, the number of polymorphic synonymous sites and non-synonymous sites, as well as the number of synonymous substitutions and non-synonymous substitutions compared to the close relative *Cardaminopsis petraea* was recorded. A Mcdonald-Kritman test of the sequence variation (Table 6.3) revealed significant departures from the neutral expectations indicating that selection was operating on the locus (Kawabe et al. 2000).

Table 6.3   McDonald-Kreitman test (McDonald and Kreitman 1991) of the sequence variation at the *PgiC* locus in *Arabidopsis thaliana*. The 2 × 2 contingency table showed significant departure from the neutral expectation that the number of polymorphic sites and substitutions should be independent of whether the site was synonymous or non-synonymous (G = 13.3, P < 0.001) (after Kawabe et al. 2000).

	Polymorphic sites in *A. thaliana*	Substitutions relative to *C. petraea*
Synonymous sites	16	55
Non-synonymous sites	15	8

The different methodologies of measuring natural selection summarised above may of course all be applied on the same trait if there is sufficient information on the genetic basis of the trait and such a combined approach will be relatively powerful.

7. Evolution of plant life history

Trade-offs and evolutionary stable strategies

If relevant genetic variation is available in a plant population, the process of natural selection will cause evolution (Darwin 1859), i.e., the physiology, morphology and life history of the plant population will adapt to the present abiotic and biotic environment. Since evolution depends on the process of natural selection, the type of evolutionary changes are limited by the mechanisms of natural selection and often, but certainly not always (Lewontin 1970), this means that the evolutionary changes should be explained at the level of the fitness of the individual plant.

A plant with a given amount of resources at a certain point in time may allocate its resources to different purposes determined by the evolved life history of the plant and impulses from the environment. For example, an annual plant will usually allocate all its resources to reproduction at the end of the growing season, whereas a perennial plant only can allocate a fraction of its resources to reproduction (e.g. Harper 1977). Some of the resources of the perennial plant need to be stored to survive during harsh periods, e.g. during a drought or a winter period. The perennial plant is said to make a trade-off between reproduction and survival, and since both features are important for the individual fitness of the plant this trade-off will be under selective pressure.

The characteristic life history of a specific plant species will to a large extent often be determined by the way natural selection has shaped the morphological and physiological features underlying the various trade-offs. Other trade-offs that are being selected to increase the fitness of the individual plant at its particular environment include the allocation between male and female sexual structures, and the allocation between structures that will increase growth, e.g. stems and leaves, and a general storage of resources (e.g. Harper 1977). For example, if the plant density is high it may be adaptive to grow rapidly due to the effect of size-asymmetric competition, but if the likelihood of a herbivore attack is high, it may be better for the plant to save resources for re-growth after a possible herbivore attack. Alternatively, it may be adaptive for the plant to invest in producing structures that reduce the likelihood of a herbivore attack, e.g. trichomes. Notice that since the likelihood and magnitude of a herbivore attack may depend on the plant density, it is not a simple task to understand and predict the evolutionary forces on the different trade-offs, which often need to be modelled in a quantitative way.

Natural selection operates by changing the frequencies of the alleles in the population as discussed in the previous chapter, and the evolutionary forces on plant life histories may be examined by making a genetic model by introducing a mutant allele that modifies the life history in a certain way. The fitness of the new mutant should now be modelled and if the fitness of the mutant is higher than the fitness of the other individuals in the population, the mutant allele will invade the population with a certain probability (6.5). However, when a complicated life history is modelled, the real challenge is often to model the fitness of the different life history strategies. In order to simplify the problem it is often assumed that the genetic system is very simple, i.e. the population is haploid, completely selfing, or asexual. In such cases the evolutionary process only depends on the frequencies of the various phenotypes in the population and not on the often-unknown genetic system behind the phenotypes (e.g. Geritz 1998).

A specific life history or phenotype may be called a pure or a singular strategy if it is a fixed strategy, as opposed to a mixed strategy where different singular strategies are chosen with a specified probability (Hofbauer and Sigmund 1998). Such a singular strategy may be an evolutionary stable strategy (ESS) if the phenotype makes a population immune to invasion by any other phenotype (Maynard Smith and Price 1973, Hofbauer and Sigmund 1998). By calculating the phenotypic fitnesses of the possible singular strategies it is possible to determine which phenotype is the ESS.

Consider a population with two singular strategies I and J in the frequencies p and $q = 1 - p$, respectively, where the fitnesses w of the strategies generally depend on the frequencies of I and J, e.g. the fitness of the I-phenotypes is $w(I, qJ + pI)$. A population consisting of I-phenotypes will be evolutionary stable if, whenever, a small amount of deviant J-phenotypes is introduced, the old I-phenotypes have a higher fitness than the J-phenotypes, i.e.,

$$w(J, qJ + pI) < w(I, qJ + pI) \tag{7.1},$$

for all sufficiently small q (Hofbauer and Sigmund 1998).

Additionally to ESS-stability, a singular strategy may show convergence stability, which ensures the gradual approach to the singular strategy by a series of small evolutionary steps (e.g. Geritz 1998). A phenotype that is convergence-stable is an evolutionary attractor in the sense that a population that starts off with a different phenotype can always be invaded by a phenotype nearer by. Unfortunately, it turns out that ESS-stability and convergence-stability are two independent stability concepts that may occur in any combination (Geritz 1998 and references therein). Thus, if an ESS is not convergence-stable, it is not an evolutionary attractor, and often any initial perturbation away from a phenotype that is not convergence-stable even tends to be amplified in the next generations (Geritz 1998).

If a phenotype is convergence-stable but not an ESS, it acts as an attractor, but in a population in which all plants have this phenotype any nearby mutant phenotype can invade the population. In simulations it has been observed how a monomorphic population of a phenotype that is convergence-stable but not an ESS diverges into two distinct phenotypes, and such singular strategies are called evolutionary branching points (Geritz 1998).

One life history trade-off, which is under continuous selective pressure, is whether a plant should produce many small seeds or fewer larger seeds. It is assumed that a plant has a fixed amount of resources allocated to reproduction, and the plant may either increase fecundity by producing more seeds or produce fewer larger seeds, which decrease seedling mortality because the seedlings have a competitive advantage (Geritz et al. 1999 and references therein). The fitness of the individual plant may be modelled by:

$$w = \frac{R}{m} f(m) \tag{7.2},$$

where R is the amount of resources allocated to reproduction, m is seed size, and $f(m)$ is the expected fecundity of a seed as a function of its size.

In a simple ecological model, where there are no spatial or density-dependent effects, the seed size that maximises fitness in (7.2) for a given shape of $f(m)$ is a convergence-stable ESS (Fig. 7.1-a). If the effects of space and size-asymmetric competition is included in the ecological model, the evolutionary dynamics become more complicated. Geritz et al. (1999) assumed that a habitat could be divided into a number of sites where a single plant can grow and that the probability of establishment at such a site was a function of the seed size. Furthermore, the seedlings were assumed to show size-asymmetric competition during the process of establishment so that larger seeds had a probability that was more than proportional to their size advantage (the process of size-asymmetric competition was modelled in a way analogous to (2.10)). They found that with increasing size-asymmetric competition an increasing number of seed sizes were convergence-stable but not ESS's, i.e., evolutionary branching points (7.1-b). Thus if sufficient size-asymmetric competition is occurring during seedling establishment we do not expect to find a single optimal seed size, instead the population should be polymorphic for different seed sizes, which also is commonly observed in natural populations (Westoby et al. 1992).

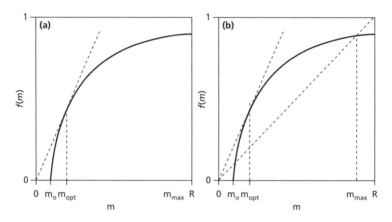

Fig 7.1 The expected fecundity of a seed as a function of its size. Below the seed size m_0 the seed is not expected to establish.

(a) In a simple ecological model, where there are no spatial or density-dependent effects, the seed size that maximises fitness, m_{opt}, is at the point where the tangent of a given $f(m)$ goes through the origin (Smith and Fretwell 1974). This seed size is a convergence-stable ESS and will act as a global evolutionary attractor (Geritz et al. 1999).

(b) In a more complicated spatial ecological model with size-asymmetric competition, there may be several evolutionary branching points at an interval between m_{opt} and m_{max}, which is the maximum seed size that ensures that the population does not go extinct (at the intersection of the main diagonal and $f(m)$). The number of evolutionary branching points increases with size-asymmetric competition until the case of complete size-asymmetric competition (larger seeds always win), where all seed sizes in the interval between m_{opt} and m_{max} are ESS's. Figure after Geritz et al. (1999).

In the above it is assumed that plants have a singular strategy of seed size, but actually many plant species seem to have a mixed strategy, where a single plant produces different types of seeds (Westoby et al. 1992). Additionally, seed size may affect seed dispersal so that there may be a 3-dimensional trade-off function between fecundity, establishment, and dispersal (Levin and Muller-Landau 2000).

The evolutionary consequences of size-asymmetric competition in a spatial setting, which due to their sedentary life form is so characteristic of plant communities (see chapter 2), is generally understudied. This is unfortunate because a number of new phenomenons seem to pop up in the evolutionary models when the individual fitness is described on the individual level in a spatial setting rather than by a mean-field-approach (e.g. Law et al. 1997). Notice how the *a priori* intuitively correct and simple result of the simple ecological model of the seed size trade-off (Fig 7.1-a) almost exploded into the complexity of a growing number of evolutionary branching points (Fig. 7.1-b) when the effects of space and size-asymmetric competition was included in the model. Perhaps many of the intuitive ecological results obtained by simple optimisation modelling will go through a similar development as described for the understanding of the evolution of seed size.

Another adaptation that independently have evolved in many plant taxa as a typical response to an arid or a ruderal habitat is the evolution of an annual life form from a perennial life form (e.g. Harper 1977, Barrett et al. 1996). The selective advantage of dying may be understood under the assumption that annual plants are able to allocate more of their resources into reproduction as they do not need resources to survive. If a perennial plant by becoming an annual may produce more than only one additional successful seed than it would produce each year as a perennial plant, it is a selective advantage to become an annual (Cole 1954, Bryant 1971). This famous result assumes that the perennial species has the same fecundity each year. If fecundity increases with age as is typically found among perennial plants, it becomes somewhat more complicated to calculate the possible selective advantage of an annual life form (Charlesworth 1994). The probability that the supposed additional seeds of annual plants are successful will depend on the abiotic and biotic environment and may be investigated experimentally using the framework developed in chapters 3 and 4.

Evolution of sex

Plants display a considerable variation in their reproductive systems, from asexual reproduction to sexual reproduction where the male and the female functions are distributed within and among plants in all possible combinations (Table 7.1).

Table 7.1 A classification of plant reproductive systems. Percentages refer to the reproductive systems recorded from the best available compilation, a survey of a large number of angiosperm species by Yampolsky and Yampolsky (1922). Gymnosperms are mostly monoecious or dioecious (from Silvertown and Charlesworth 2001).

Description	Botanical term and definition	Occurrence in plants and examples
Asexual	*Apomictic*: Seeds have the same genotype as their mother. In angiosperms pollination and fertilization of the endosperm (pseudogamy) is common.	Many *Taraxacum* spp.
Sexually monomorphic	*Hermaphrodite*: Flowers have both male and female functions.	(72%) e.g. most rose cultivars, *Rosa* spp.
	Monoecious: Separate sex flowers on the same individual plants.	(5%) e.g. cucumbers
	Gynomonoecious: Both female (male-sterile) and hermaphrodite flowers occur on the same individuals.	(2.8%) many Asteraceae including *Bellis* and *Solidago*
	Andromonoecious: Both male (female-sterile) and hermaphrodite flowers occur on the same individuals.	(1.7%) Most Apiaceae, e.g. carrot *Daucus carota*
Sexually polymorphic	*Dioecious*: Separate sex (male and female) individuals.	(4%) e.g. hollies, *Ilex* spp.
	Gynodioecious: Individuals either female or hermaphrodite.	(7%) many Lamiaceae, e.g. Glechoma, *Thymus* spp.
	Androdioecious: Individuals either male or hermaphrodite.	Very rare. E.g. *Mercurialis annua*

It is noteworthy that most plants are sexual and the explanation of this phenomenon has since long been an evolutionary puzzle. It turns out that there is a multitude of non-exclusive ecologically and genetically based hypotheses that explain the evolution of sexual systems (e.g. Maynard Smith 1978, Barton and Charlesworth 1998, Holsinger 2000).

First of all, there is a cost associated with producing males or a male function. Everything else being equal, a population of asexual plants is expected to have twice as high intrinsic population growth rate as a dioecious population, because the males in the dioecious population do not produce any offspring (e.g. Maynard Smith 1978). Furthermore, some of the female ovules may not be fertilised due to an insufficient number of pollen, even though a considerable amount of resources may be used to attract pollinating insects. In dioecious or self-incompatible hermaph-roditic plants species the lack of suitable pollen donors during a bot-tleneck or a colonisation event may even lead to local extinction (Baker 1955). Most plants are hermaphroditic and such a breeding system may be regarded as a relatively inexpensive way of having a sexual repro-ductive system, however, the male function may reduce the potential fecundity of the hermaphrodites considerably as seen in gynodioecious species (Fig. 7.2).

The infamous two-fold advantage of asexual plants hinges on a number of important assumptions, and most importantly that the phe-notypes of the asexual and sexual populations, except from the sexual function, are equal. However, the mating system is known to influence the genetic variation in the population (see chapter 5) and consequently the phenotypes. Even a small genetic variation between the asexual and sexual phenotype may be of critical importance in connection with the other assumptions in the arguments leading to the notion of a two-fold advantage of asexual plants. For example:

1) When populations reach carrying capacity, a high intrinsic growth rate may not be equally important in determining the fate of the asexual population if the sexual population has a superior competi-tive ability or a higher carrying capacity (Doncaster et al. 2000).
2) *The tangled bank*: if the habitat is comprised of a number of different ecological niches then the number of available niches of an asexual population may be restrained by the lack of genetic variation (Bell 1982, Lomnicki 2001).
3) *The Red Queen*: if the environment is changing rapidly with time, e.g. by continuos co-evolution of new resistant parasites, the sexual population may be better at adapting to the changing environment (e.g. Crow 1992, Hamilton 1993).
4) Theoretically, it has been shown that modifiers that increase the frequency of sexual reproduction or recombination tend to become associated with positively selected alleles, and consequently will increase in frequency (e.g. Barton and Charlesworth 1998 and refer-ences therein).

In addition to an expected reduction in the intrinsic population growth rate, sexual reproduction greatly increases the likelihood of evo-lutionary conflicts, which often have a detrimental effect on the group fitness. In an asexual population all the genes present in an individual are in permanent association and share their evolutionary fate. That is, the fitness effects of one allele on the asexual plant affects its own transmission in the same way as that of all the other alleles in the plant. In contrast, in a sexual population the associations among alleles at different loci are temporary and are broken up by sex and recombination. Intragenomic conflicts are therefore more likely among sexual plants (e.g. Partridge and Hurst 1998) and may be included as a potential long-term cost of sexual reproduction. One example of an evolutionary conflict in sexual plants is cytoplasmic inherited male-sterility, which often is the underlying cause of a gynodioecious reproductive system (Couvet et al. 1990). The cytoplasmic genes, e.g. in the mitochondria, are only rarely transmitted through pollen and the fitness of a mitochondrial gene is unaffected by the paternal fitness of a plant. Consequently, from a mitochondrial genes point of view it would be better if all the resources used to produce the male functions was allocated toward a higher fecundity, and any increase in fecundity caused by inducing male sterility would be an adaptive advantage (Fig 7.2 curve 1). However, male-sterility may also be an advantage for the whole individual and in particular for a nuclear inherited gene if the increase in fecundity more than compensates for the loss in paternal fitness (Lewis 1941, Fig 7.2 curve 2).

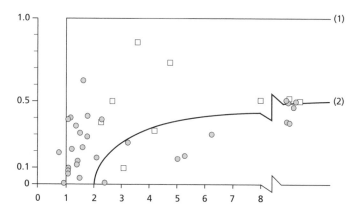

Fig 7.2   Relationship between frequencies of females and their relative fecundity: theoretical expectations and observations. Curve 1 is the expected minimum relative fecundity of females in the case of cytoplasmic inheritance of male-sterility. Curve 2 is the expected minimum relative fecundity of females in the case of nuclear inheritance of male-sterility in an outcrossing population (Lewis 1941, Charlesworth and Charlesworth 1978). The dots represent observations of the frequencies of females in different populations of gynodioecious species based on seed set and the squares represent observations that incorporate additional fitness components. Figure after Couvet et al. (1990).

One beneficial long-term effect of sex and recombination is that the rearrangement of the alleles may increase the variation in fitness either by association of deleterious alleles, which may be purged from the population, or by association of advantageous alleles, which may be positively selected (e.g. Barton and Charlesworth 1998 and references therein). The increased variation in fitness in sexual populations may through natural selection in the long run lead to superior adapted phenotypes with less genetic load. Such a superior group fitness of sexual plants may explain the observed high frequency of sexual plant species. Even though asexual mutants may have an individual fitness advantage and may spread in some environments (Bell 1982), asexual plant species may face a higher risk of extinction (Maynard Smith 1978, Holsinger 2000).

Evolution of the selfing rate

Among the sexual plants the process of mating may take many different and interesting forms as briefly shown in Table 7.1, and the relationship between the flower morphology and the mating system has inspired numerous evolutionary biologists in their thinking.

One of the most studied features of the mating system is the evolution of the selfing rate, and especially the explanation of the observed apparently evolutionary stable strategy of partly selfing mating systems (Fig. 5.2). Since long, it has been observed that inbreed individuals have reduced viability and fecundity (inbreeding depression), and Darwin (1877) proposed that the diversity of floral forms could be explained as a result of natural selection to ensure cross-pollination. Fisher (1941) pointed out that an allele that promoted selfing would sweep through an outcrossing population, unless opposed by a strong selective force due to the transmission advantage of self-fertilising plants (a self-fertilising plant has two gamete copies in each seed, whereas an outcrossing plant has only one). Since then, the evolution of the selfing rate has been explained mainly as controlled by the two opposing selective forces of the transmission advantage of self-fertilising plants and inbreeding depression (e.g. Feldman and Christiansen 1984, Lande and Schemske 1985, Charlesworth and Charlesworth 1987, Charlesworth et al. 1990, Uyenoyama and Waller 1991a, b, c, Damgaard et al. 1992, Damgaard 1996, Cheptou and Schoen 2002).

The genetic basis of inbreeding depression is only partly known, but thought to be a mixture of partly recessive deleterious alleles and overdominant loci (e.g. Charlesworth and Charlesworth 1999). Inbreeding has the effect of purging the recessive deleterious alleles from the population because the alleles are "exposed" relatively more often in homozygous individuals, but it is outside the scope of this monograph to review the complicated interplay between the selfing rate and the genetic basis of the inbreeding depression. The understanding of the effect of inbreeding depression is further complicated by the fact that the amount of inbreeding depression depends on e.g. the mating history as well as the biotic and abiotic environment (e.g. Damgaard et al. 1992, Damgaard and Loeschcke 1994b, Cheptou and Schoen 2002).

The matter may be muddled even further by noting that inbreeding depression is only one of the processes that may affect the selfing rate (Fig. 7.3). Other selective processes such as segregation distortion, pollen competition, and selective abortion may also influence the selfing rate (Uyenoyama et al. 1993, Holsinger 1996). Additionally, a trivial but often-overlooked fact is that the process of pollen transfer itself predisposes the selfing rate in the absence of any selective forces (Gregorius et al. 1987, Holsinger 1991, Uyenoyama et al. 1993, Damgaard and Loeschcke 1994a, Damgaard and Abbott 1995, Holsinger 1996). That is, the individual selfing rate depends, at least in part, on the relative amount of self- and outcross pollen found on the stigmas of the plant. Since the distribution of self- and outcross pollen depends on the density- and frequency dependent processes of pollen transfer, the selfing rate of a genotype both depends on the density and frequency of the genotype. Furthermore, since the reproductive success of a partially self-fertilising plant depends on the combined effect of producing either outcrossed or selfed seeds and siring seed on other plants, the reproductive success of different genotypes are also density- and frequency dependent (e.g. Holsinger 1996).

Various other ecological factors may also affect the evolution of selfing rate. For example, if foreign pollen are limited, e.g. after a colonisation event, then the possibility of self-fertilising may be an adap-

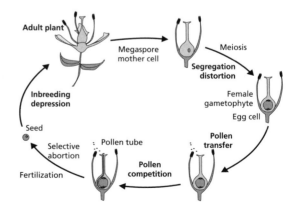

Fig. 7.3 Typical phases in the mating system of a hermaphroditic plant. Heritable differences among individuals in characteristics that affect any of the life cycle transitions may lead to an evolutionary response. Some of the processes that affect each of these transitions are identified in bold face on the figure.

Figure after Holsinger (1996).

tive strategy (Baker 1955). It is clear that with such an overwhelming number of interacting genetic and ecological factors, which control the evolution of the selfing rate, it is difficult to give general answers to whether a partly selfing mating system is evolutionary stable. Thus it is necessary to examine each species individually in the assessment of the relative evolutionary importance of the different factors.

The allocation of resources between the male and the female gender of hermaphroditic plants is effected by average selfing rate in the population. It has been observed that self-fertilising plants have fewer pollen per ovules (e.g. Cruden and Lyon 1985) and this phenomenon has been explained by the fact that on average a population of highly selfing plants sire fewer seeds than a population of outcrossing plants. This means that for the individual highly self-fertilising plant there is a relatively higher fitness return when resources are invested in the female gender (Charlesworth and Charlesworth 1981, Charnov 1982). The allocation towards the female gender in a self-fertilising population will go as far as the marginal returns to additional pollen productions are equal to those of additional ovule production (Charnov 1982). The altered sex allocation in a partly self-fertilising plant population and the phenomenon of inbreeding depression may again affect the possible evolution of a hermaphroditic mating system towards either gynodioecy or dioecy as shown in Fig. 7.2 (Charlesworth and Charlesworth 1981).

Speciation
The understanding of the diversity of life and consequently the process of speciation has always been at the core of the modern biological science. Species are man-made sets of elements constructed with the purpose of facilitating the discussion and thinking on the diversity of life, and the concept of a species is more abstract than the previously examined concepts of a population, a habitat, or a geographical region (e.g. Otte and Endler 1989). The best known definition of a species is the "biological" species concept by Mayr (1942), where species are regarded as groups of interbreeding populations which are reproductively isolated. However, especially for plants this definition has caused some problems due to e.g. self-fertilisation and polyploidy (e.g. Otte and Endler 1989), and it is probably better to follow Stebbins (1950): "The wisest course would seem to avoid defining species too precisely and to be tolerant of somewhat different species concepts held by other workers. The one principle which is unavoidable is that species are based on discontinuities in the genetic basis of the variation pattern rather than on the amount of difference in their external appearance between extreme or even typical individual variants".

Traditionally, the gradual build up of reproductive isolation between geographical isolated populations has been regarded as the crucial step in speciation later followed by differential adaptive selection (allopatric speciation, Mayr 1942). The genetic background of the reproductive isolation among plant species is now beginning to be understood in some detail. For example, in an intraspecific cross between two rice varieties (Indica: Kasalath and Japonica: Nipponbare) the deviations from the expected Mendelian segregation ratios were studied at a number of genetic markers (Harushima et al. 2001). They found that a total of 33 chromosome segments contributed to reproductive isolation between the two varieties and approximately half of these chromosome segments affected the transmission rate through the gametophyte (Harushima et al. 2001).

Lately more attention has been paid to possible ecological factors that may lead to speciation, and it has been suggested that natural selection arising from differences between environments and competition for resources may lead to rapid speciation (Schluter 2000). The most recent speciation and diversification events, which consequently are the most clear examples of the speciation process, have occurred on new isolated islands where a number of niches have been available for the first taxa that arrived at the island. For example, a common ancestor of the Hawaiian silversword alliance (Agyroxiphium and others) colonised the Hawaiian archipelago early after it emerged from the sea about 5 million years ago. Since then an adaptive radiation has taken place so that now 28 species exploit a huge array of habitat types including cold arid alpine regions, hot exposed cinder cones, wet bogs where water uptake is nevertheless inhibited, and dimly lit rain forest understories (Schluter 2000 and references therein).

Additionally, it has been demonstrated theoretically that speciation may occur at evolutionary branching points, e.g. as a result of size-asymmetric competition, when assortative mating is taking place (Dieckmann and Doebeli 1999, Doebeli and Dieckmann 2000). Thus, sympatric speciation, where new species arise without geographical isolation, which previously have been thought to be unlikely due to gene flow between the emerging species, is now discussed as a probable mode of speciation.

As repeatedly concluded in the above sections, i.e. that if both ecological and genetic factors play a role in the evolutionary processes, it must also be concluded that the process of speciation depends on both ecological and genetic factors. Consequently, the study of both micro- and macro evolutionary processes will need to integrate both ecological and genetic factors in both experimental and theoretical work in order to understand the evolutionary processes at a satisfactory level.

Appendix A

Parameters and variables with a fixed usage in chapters 2 – 4

Parameter	Description
v	Plant size
w	Plant size at the end of the growing season
v_m	Plant size of an isolated plant
w_m	Plant size at the end of the growing season of an isolated plant
κ	Initial growth rate constant
δ	Parameter that determine the location of the inflection point in the Richards growth model
a	Degree of size-asymmetric growth
n	Number of interacting plants
x	Density of plants
x_0	Density of seeds
$\alpha,\ \beta,\ \phi,\ \theta$	Shape parameters in plant size (fecundity) – density response models
m_d	Density-dependent mortality parameter
m_i	Density independent probability of germination and establishment
p	Probability of germination, establishment and reaching reproductive age
b	Fecundity (number of seeds produced)
U	Competitive neighbourhood function
r	Distance
D	Distribution of dispersal distances
$C(r)$	Spatial covariance at distance r
$\bar{C}$	Average spatial covariance
c_{ij}	Per capita competition coefficient of species j on i
g	Generation

Appendix B Nonlinear regression

Traditionally, biologists have relied on linear models, e.g. linear regression, ANOVA etc., in the statistical analysis of biological effect studies. However, if we want to understand the biological processes by gathering ecological data and modelling the ecological dynamic processes underlying the data, as exemplified in this monograph, we are forced to use nonlinear models in a statistical way.

A linear model is a model where the parameters are linear. For example, the effect of plant density on the average plant size as described by model (2.13), $v(x) = v_m/(1 + \beta x^\phi)$, is a nonlinear model because the parameters β and ϕ cannot be separated by a plus or a minus sign. The model predicts the expected plant size, $v(x)$, based on the plant density, x, given the three parameters (v_m, β, ϕ) and each observed value in a data set may now be compared with its corresponding expected value with the same density.

Generally, a nonlinear model may be fitted to data, or put in another way, the parameters may be estimated from a data set, using the likelihood approach (Edwards 1972, Seber and Wild 1989). The likelihood function, $p(Y|\theta)$, is the probability of observing the data (denoted by Y) given (or conditional on) a specific model (denoted by the model parameters θ, that may be a vector if there are more parameters in the model). The likelihood function makes the mathematical connection between the parameters in the model and the data. In order to formulate the likelihood function the residual variation has to be specified, and here we will assume that the residuals (observed value – expected value, i.e., $y_i - f(x_i, \theta)$) are independent and identically normal distributed with zero mean and a variance of σ^2. Then the likelihood function will be:

$$p(Y|\theta, \sigma^2) = (2\pi\sigma^2)^{-\frac{n}{2}} \exp\left(-\left(\frac{1}{2} \prod_{i=1}^{n} \frac{(y_i - f(x_i, \theta))^2}{\sigma^2}\right)\right) \tag{B.1},$$

where i signifies observation i out of n observations.

One of the immediate goals of the regression is to obtain the maximum likelihood estimate of the parameter, $\hat{\theta}$, in the model. The maximum likelihood estimate is the parameter value for which the likelihood function is maximised. The maximisation of the likelihood function is usually done numerically and may be done using different software packages including spreadsheets.

Often it is easier to calculate the logarithm of the likelihood function, the log-likelihood function:

$$l(Y \mid \theta, \sigma^2) = -\frac{n}{2} \log(2\pi\sigma^2) - \frac{1}{2\sigma^2} S(\theta) \tag{B.2},$$

where $S(\theta) = \sum_{i=1}^{n} (y_i - f(x_i, \theta))^2$ is the sum of the squared residuals.

Given the model parameters, the maximum likelihood estimate of the variance is equal to: $\sigma^2 = S(\theta)/n$, and inserting this into the log-likelihood function (B.2) the log-likelihood function that needs to be maximised is simply (Seber and Wild 1989):

$$l(Y \mid \theta) = -\frac{n}{2} \log(2 \; S(\theta)/n) - \frac{n}{2} \tag{B.3}.$$

The assumption of identically distributed residuals is often violated and it is, in those cases, necessary to transform the observed and expected values in order to homogenise the variance of the residuals. One quite general way of homogenising the variance of a continuos character is by using the Box- Cox transformation (Seber and Wild 1989):

$$y^{(\lambda)} = \begin{cases} \dfrac{(y + \lambda_2)^{\lambda_1}}{\lambda_1} \; , & \lambda_1 \neq 0 \\ \log(y + \lambda_2) \; , & \lambda_1 = 0 \end{cases} \tag{B.4},$$

for $y > \lambda_2$. The Box-Cox transformation is a generalisation of e.g. performing no transformation ($\lambda_1 = 1$), the square-root transformation ($\lambda_1 = 1/2$), and the log-transformation ($\lambda_1 = 0$). The appropriate values of (λ_1, λ_2) may be found by assuring that $y > -\lambda_2$ for all observed data, and by maximising the following log-likelihood function with respect to λ_1.

$$l(Y \mid \theta, \lambda_1) = -\frac{n}{2} \log(2\pi S^{(\lambda)}(\theta)/n) - \frac{n}{2} + (\lambda_1 - 1)\sum_{i=1}^{n} \log(y_i + \lambda_2) \tag{B.5},$$

where (Seber and Wild 1989). When the right $S^{(\lambda)}(\theta) = \sum_{i=1}^{n}(y_i^{(\lambda)} - f(x_i, \theta)^{(\lambda)})^2$

transformation has been found and the residuals checked for approximate normality, the maximum likelihood estimates of the model parameters may be found.

In some cases it may be beneficial to try to simplify the model by setting a parameter to a fixed value. For example, in model (2.13), $v(x) = v_m/(1 + \beta x^\phi)$, the functional relationship between density and size may be simplified considerably if the parameter ϕ, which is a measure of the rate at which competition decays as a function of distance between plants, is set to one. It can be tested whether the data supports a more complicated model be comparing the maximum values of the log-likelihood functions of the reduced model ($H_0 : \theta_r = \theta_{r0}$) and the full model ($H_1 : \theta_r \neq \theta_{r0}$). The test value is twice the difference between the maximised log-likelihood functions, and for n large, this test value is asymptotically chi-square-distributed, i.e.,

$$2(l(Y \mid \hat{\theta}_{H_1}) - l(Y \mid \hat{\theta}_{H_0})) \approx \chi^2 (d_{H_1} - d_{H_0}) \tag{B.6},$$

where $d_{H1} - d_{H0}$, the degrees of freedom in the chi-square-distribution, is the reduction in the number of free parameters from the full model to the reduced model (Stuart and Ord 1991).

A *Mathematica* notebook exemplifying a linear regression procedure may be downloaded from my webpage. If you do not have access to *Mathematica*, a *Mathematica Reader* may be downloaded from www.wolfram.com.

Appendix C   Bayesian inference

The view of statistics employed in this monograph is a pragmatic one, and there will be no review of the formal discussion of the differences between the traditional "frequentist" approach and the Bayesian approach (e.g. Edwards 1972, Carlin and Louis 1996). Here, it suffices to mention that in the Bayesian approach the model parameters are treated as random variables from an unknown distribution rather than having a true but unknown value.

I would argue that the Bayesian view of a model parameter corresponds to the way a "typical" ecologist thinks and how he plans an experiment. Does it give much meaning to talk about the *true* size of an isolated plant in model (2.13)? Almost every ecologist would think of the parameter: "the size of an isolated plant" as a random variable from an unknown distribution. Furthermore, in some applied ecological questions, e.g., risk assessment of genetically modified plants and the management of natural habitats, it is desirable to be able to predict the probability of a specific ecological scenario and as demonstrated in example 4.2, the Bayesian statistical approach is ideally suited for this purpose.

The Bayesian view of model parameters as distributions of random variables has the mathematical consequence that it is necessary to specify a *prior distribution* of the model parameter. That is, the experimenter's knowledge of the likely values of the parameter *before* the experiment is conducted has to be described mathematically. For example, if the experiment has been conducted previously, the previously obtained results may be used as a prior distribution. If no prior knowledge on the model parameter is available, an uninformative prior distribution may be chosen, in which case the mathematics of the Bayesian approach becomes simpler and in fact almost identical to the traditional likelihood approach! Actually, it is rare that no prior knowledge on the model parameter exists, but in a predominately "frequentist" scientific world it may be strategic to assume an uninformative prior distribution anyway.

Bayes' theorem defines the *posterior distribution* of the model parameter, which tells us what we know about θ given the knowledge of the data Y, as:

$$p(\theta|Y) = \frac{p(Y|\theta)\,p(\theta)}{\int p(Y|\theta)\,p(\theta)\,d\theta} = c\,p(Y|\theta)p(\theta) \propto p(Y|\theta)p(\theta) \tag{C.1},$$

where $p(Y \mid \theta)$ is the likelihood function (see B.1), $p(\theta)$ is the prior distribution of the model parameter, and the integral may be regarded as a normalising constant, $c^{-1} = \int p(Y \mid \theta) \, p(\theta) \, d\theta$ (Box and Tiao 1973). If we assume an uniform uninformative prior distribution, then the posterior distribution is proportional to the likelihood function, i.e. $p(\theta \mid Y) \propto p(Y \mid \theta)$.

One of the strengths of the Bayesian approach is that it provides a mathematical formulation of how previous knowledge may be combined with new knowledge. Indeed sequential use of Bayes' theorem allows us to continually update information about a parameter θ as more observations are made. Suppose we have an initial sample of observations, Y_1, then the posterior distribution of θ is:

$$p(\theta \mid Y_1) \propto p(\theta) \, p(Y_1 \mid \theta) \tag{C.2},$$

then we make another sample of observations, Y_2, and now the posterior distribution of is:

$$p(\theta \mid Y_1, Y_2) \propto p(\theta) \, p(Y_1 \mid \theta) \, p(Y_2 \mid \theta) \propto p(\theta \mid Y_1) \, p(Y_2 \mid \theta) \tag{C.3},$$

thus, the combined knowledge of θ after two experiments, i.e. the posterior distribution $p(\theta \mid Y_1, Y_2)$, may simply be found by multiplying the prior distribution of from before the first experiment with the two likelihood functions of the two experiments, which is equivalent to multiplying the posterior distribution of θ after the first experiment with the likelihood function of the second experiment. A similar procedure may be followed when the combined knowledge of any number of different experiments needs to be established (Box and Tiao 1973).

Since the Bayesian approach treat model parameters as random variables from an unknown distribution rather than having an unknown true value, the concept of a confidence interval, in which the true value with a specified probability has to be found, is meaningless. Instead, a credibility interval is defined as the interval where a certain proportion of the probability density of the posterior distribution is found. For example, a 95% *credibility interval* is the interval between the 2.5% percentile and the 97.5% percentile of the posterior distribution.

The testing of different hypotheses may, of course, also be done in the Bayesian approach (Carlin and Louis 1996). However, the Bayesian testing procedure using Bayes' factors is less intuitive as the other Bayesian concepts and has not been adopted by the scientific ecological community. Therefore, as statistics also is about communicating your experimental results in a readily understandable way, it is in my opinion better to rely on the well-known likelihood ratio tests (B.6) or use the intuitive testing procedure of comparing different credibility intervals (see example 2.1).

Traditionally, the main disadvantage of the Bayesian approach was that large numerical computations are required, but with the use of modern computers this problem has disappeared. Furthermore, relatively user-friendly software like WinBugs (http://www.mrc-bsu.cam.ac.uk/bugs/) has been developed for building and analysing stochastic Bayesian models.

A *Mathematica* notebook exemplifying a Bayesian approach to data analysis may be downloaded from my webpage. If you do not have access to *Mathematica*, a *Mathematica Reader* may be downloaded from www.wolfram.com.

Appendix D Stability of discrete dynamic systems

Consider a system of recursive equations of the form:

$$\begin{cases} x_1\,(t+1) = f_1\,(x_1(t),\, x_2(t)) \\ x_2\,(t+1) = f_2\,(x_1(t),\, x_2(t)) \end{cases} \tag{D.1}$$

Here we will only consider a system of two recursive equations, but the methodology is readily generalised to n recursive equations (e.g. Elaydi 1999).

The equations may be solved:

$$\{\hat{x}_1,\, \hat{x}_2\} = \{x_1(t+1),\, x_2(t+1)\} = \{x_1(t),\, x_2(t)\} \tag{D.2}$$

and the possible local stability of an equilibrium solution (D.2) may be evaluated by a linearisation technique: The nonlinear system of recursive equations is linearised around the equilibrium solution by a Taylor's series expansion of (D.1) at (D.2):

$$\begin{cases} x_1(t+1) = f_1(x_1(t), x_2(t)) \approx \hat{x}_1 + \dfrac{\partial f_1(\hat{x}_1, \hat{x}_2)}{\partial x_1}\,(x_1(t) - \hat{x}_1) + \dfrac{\partial f_1(\hat{x}_1, \hat{x}_2)}{\partial x_2}\,(x_2(t) - \hat{x}_2) \\[4mm] x_2(t+1) = f_2(x_1(t), x_2(t)) \approx \hat{x}_2 + \dfrac{\partial f_2(\hat{x}_1, \hat{x}_2)}{\partial x_1}\,(x_1(t) - \hat{x}_1) + \dfrac{\partial f_2(\hat{x}_1, \hat{x}_2)}{\partial x_2}\,(x_2(t) - \hat{x}_2) \end{cases} \tag{D.3}$$

The Taylor's series expansion (D.3) suggest that the local stability of the solution (D.2) may be investigated by the corresponding linear homogenous system with the Jacobian matrix:

$$\begin{bmatrix} \dfrac{\partial f_1(\hat{x}_1, \hat{x}_2)}{\partial x_1} & \dfrac{\partial f_1(\hat{x}_1, \hat{x}_2)}{\partial x_2} \\[4mm] \dfrac{\partial f_2(\hat{x}_1, \hat{x}_2)}{\partial x_1} & \dfrac{\partial f_2(\hat{x}_1, \hat{x}_2)}{\partial x_2} \end{bmatrix} \tag{D.4}$$

and it is known that a linear homogenous system with the Jacobian matrix (D.4) is stable if the absolute values of the eigenvalues are less than one (e.g. Elaydi 1999).

A *Mathematica* notebook exemplifying the local stability analysis of a system of recursive equations may be downloaded from my webpage. If you do not have access to *Mathematica*, a *Mathematica Reader* may be downloaded from www.wolfram.com.

References

Abbott, R. J., and M. F. Gomes. 1989. Population genetic structure and outcrossing rate of *Arabidopsis thaliana*. Heredity **62**:411-418.

Abrams, P. A. 1996. Evolution and the consequences of species introduction and deletions. Ecology **77**:1321-1328.

Adler, R. A., and J. Mosquera. 2000. Is space necessary? Interference competition and limits to biodiversity. Ecology **81**:3226-3232.

Allard, R. W., S. K. Jain, and P. L. Workman. 1968. The genetics of inbreeding populations. Advances in Genetics **14**:55-131.

Antonovics, J., and N. L. Fowler. 1985. Analysis of frequency and density effects on growth in mixtures of *Salvia splendens* and *Linum grandiflorum* using hexagonal fan designs. Journal of Ecology **73**:219-234.

Asmussen, M. A. 1983a. Density-dependent selection incorporating intraspecific competition. I. A haploid model. Journal of Theoretical Biology **101**:113-127.

Asmussen, M. A. 1983b. Density-dependent selection incorporating intraspecific competition. II. A diploid model. Genetics **103**:335-350.

Asmussen, M. A., and E. Basnayake. 1990. Frequency-dependent selection: The high potential for permanent genetic variation in the diallelic, pairwise interaction model. Genetics **125**:215-230.

Baker, H. G. 1955. Self-compatibility and establishment after long-distance dispersal. Evolution **9**:347-349.

Balding, D. J., M. Bishop, and C. Cannings, editors. 2001. Handbook of statistical genetics. Wiley, Chichester.

Barrett, S. C. H., L. D. Harder, and A. C. Worley. 1996. Comparative biology of plant reproductive traits. Philosophical Transactions of the Royal Society, London Series B **351**:1272-1280.

Barton, N. H., and B. Charlesworth. 1998. Why sex and recombination. Science **281**:1986-1989.

Barton, N. H., and M. Turelli. 1991. Natural and sexual selection selection on many loci. Genetics **127**:229-255.

Beerli, P., and J. Felsenstein. 2001. Maximum likelihood estimation of a migration matrix and effective population sizes in n subpopulations by using a coalescent approach. Proceedings of the National Academy of Science **98**:4563-4568.

Begon, M. 1984. Density and individual fitness: asymmetric competition. Pages 175-194 *in* B. Shorrocks, editor. Evolutionary Ecology. Blackwell, Oxford.

Bell, G. 1982. The masterpiece of nature: the evolution and genetics of sexuality. University of California Press, Berkeley.

Bennett, J. H., and F. E. Binet. 1956. Association between mendelian factors with mixed selfing and random mating. Heredity 10:51-56.

Berger, U., and H. Hildenbrandt. 2000. A new approach to spatially explicit modelling of forest dynamics: spacing, ageing and neighgourhood competition of mangrove trees. Ecological Modelling 132:287-302.

Berkowicz, R., H. R. Olesen, and U. Torp. 1986. The Danish Gaussian air pollution model (OML): Description, test and sensitivity analysis in view of regulatory applications. in V. C. de Wispelaere, F. A. Schiermeier, and N. V. Gillani, editors. Air Pollution Modeling and its Application. Plenum Press, New York.

Bleasdale, J. K. A., and J. A. Nelder. 1960. Plant populations and crop yield. Nature 188:342.

Bodmer, W. F. 1965. Differential fertility in population genetics models. Genetics 51:411-424.

Bolker, B. M. 2003. Combining endogenous and exogenous spatial variability in analytical population models. Theoretical Population Biology 64:255-270.

Bolker, B. M., D. H. Deutschman, G. Hartvigsen, and D. L. Smith. 1997. Individual-based modelling: what is the difference? Trends in Ecology and Evolution 12:111.

Bolker, B. M., and S. W. Pacala. 1999. Spatial moment equations for plant competition: Understanding spatial strategies and the advantages of short dispersal. American Naturalist 153:575-602.

Bolker, B. M., S. W. Pacala, and S. A. Levin. 2000. Moment methods for ecological processes in continuous space. Pages 388-411 in U. Dieckmann, R. Law, and J. A. J. Metz, editors. The geometry of ecological interactions: Simplifying spatial complexity. Cambridge University Press, Cambridge.

Box, G. E. P., and G. C. Tiao. 1973. Bayesian inference in statistical analysis. Addison-Wesley, Reading.

Bryant, E. H. 1971. Life history consequences of natural selection: Cole's result. American Naturalist 105:75-76.

Bullock, J. M., and R. T. Clarke. 2000. Long distance seed dispersal by wind: measuring and modelling the tail of the curve. Oecologia 124:506-521.

Bundgaard, J., and F. B. Christiansen. 1972. Dynamics of populations. I. Selection components in an experimental population of Drosophila melanogaster. Genetics 71:439-460.

Bürger, R. 2000. The mathematical theory of selection, recombination, and mutation. Wiley, Chichester.

Caballero, A., A. M. Etheridge, and W. G. Hill. 1992. The time of detection of recessive visible genes with non-random mating. Genetical Resource, Cambridge **62**:201-207.

Caballero, A., and W. G. Hill. 1992a. Effective size of nonrandom mating populations. Genetics **130**:909-916.

Caballero, A., and W. G. Hill. 1992b. Effects of partial inbreeding on fixation rates and variation of mutant genes. Genetics **131**:493-507.

Callaway, R. M. 1995. Positive interactions among plants. The Botanical Review **61**:306-349.

Campbell, D. R. 1989. Measurements of selection in a hermaphroditic plant: variation in male and female pollination success. Evolution **43**:318-334.

Carlin, B. P., and T. A. Louis. 1996. Bayes and empirical Bayes methods for data analysis. Chapman & Hall, London.

Caswell, H. 2001. Matrix population models: Construction, analysis, and interpretation, 2nd edition. Sinauer, Sunderland.

Cernusca, A., U. Tappeiner, and N. Bayfield, editors. 1999. Land-use changes in European ecosystems: ECOMONT – concepts and results. Blackwell Wissenshafts-Verlag, Berlin.

Charlesworth, B. 1992. Evolutionary rates in partially self-fertilizing species. American Naturalist **140**:126-148.

Charlesworth, B. 1994. Evolution in age-structured populations, 2 edition. Cambridge University Press, Cambridge.

Charlesworth, B., and D. Charlesworth. 1978. A model for the evolution of dioecy and gynodioecy. American Naturalist **112**:975-997.

Charlesworth, B., and D. Charlesworth. 1999. The genetic basis of inbreeding depression. Genetical Resource, Cambridge **74**:329-340.

Charlesworth, D., and B. Charlesworth. 1981. Allocation of resources to male and female functions in hermaphrodites. Biological Journal of the Linnean Society **15**:57-74.

Charlesworth, D., and B. Charlesworth. 1987. Inbredding depression and its evolutionary consequences. Annual Review of Ecology and Systematics **18**:237-268.

Charlesworth, D., M. T. Morgan, and B. Charlesworth. 1990. Inbreeding depression, genetic load, and the evolution of outcrossing rates in a multilocus system with no linkage. Evolution **44**:1469-1489.

Charnov, E. L. 1982. The theory of sex allocation. Princeton University Press, Princeton.

Cheptou, P.-O., and D. J. Schoen. 2002. The cost of fluctuating inbreeding depression. Evolution **56**:1059-1062.

Chesson, P. 2003. Understanding the role of environmental variation in population and community dynamics. Theoretical Population Biology **64**:253-254.

Christiansen, F. B. 1999. Population genetics of multiple loci. Wiley, Chichester.

Christiansen, F. B. 2002. Density dependent selection. in R. Singh, S. K. Jain, and M. Uyenoyama, editors. The evolution of population biology: modern synthesis.

Christiansen, F. B., V. Andreasen, and E. T. Poulsen. 1995. Genotypic proportions in hybrid zones. Journal of Mathematical Biology **33**:225-249.

Christiansen, F. B., and T. M. Fenchel. 1977. Theories of populations in biological communities. Springer-Verlag, Berlin.

Christiansen, F. B., and V. Loeschcke. 1990. Evolution and competition. Pages 367-394 *in* K. Wöhrmann and S. K. Jain, editors. Population biology. Springer, Berlin.

Clark, J. S., and O. N. Bjørnstad. 2004b. Population time series: process variability, observation errors, missing values, lags, and hidden states. Ecology.

Clauss, M. J., and L. W. Aarssen. 1994. Phenotypic plasticity of size-fecundity relationships in *Arabidopsis thaliana*. Journal of Ecology **82**:447-455.

Cockerham, C. C., P. M. Borrows, S. S. Young, and T. Prout. 1972. Frequency-dependent selection in randomly mating populations. American Naturalist **106**:493-515.

Cole, L. C. 1954. The population consequences of life history phenomena. Quarterly Review of Biology **29**:103-137.

Conner, E. F., and D. Simberloff. 1979. The assembly of species communities: chance or competition. Ecology **60**:1132-1140.

Connolly, J. 1986. On difficulties with replacement-series methodology in mixture experiments. Journal of Applied Ecology **23**:125-137.

Coomes, D. A., M. Rees, L. Turnbull, and S. Ratcliffe. 2002. On the mechanisms of coexistence among annual-plant species, using neighbourhood techniques and simulation models. Plant Ecology **163**:23-38.

Cousens, R. 1991. Aspects of the design and interpretation of competition experiments. Weed Technology **5**:664-673.

Cousens, R. 2001. My view. Weed Science **49**:579-580.

Couvet, D., A. Atlan, E. Belhassen, C. Gliddon, P. H. Gouyon, and F. Kjellberg. 1990. Co-evolution between two symbionts: the case of cytoplasmic male-sterility in higher plants. Pages 225-249 *in* D. Futuyma and J. Antonovics, editors. Oxford Surveys in Evolutionary Biology. Oxford University Press, NewYork.

Crawford, T. J. 1984. What is a population? Pages 135-173 *in* B. Shorrocks, editor. Evolutionary Ecology. Blackwell, Oxford.

Crawley, M., R. S. Hails, M. Rees, D. Kohn, and J. Buxton. 1993. Ecology of transgenic oilseed rape in natural habitats. Nature **363**:620-622.

Crawley, M., and S. W. Pacala. 1991. Herbivores, plant parasites, and plant diversity. *In* C. A. Toft, A. Aeschlimann, and L. Bolis, editors. Parasite-host associations – coexistence or conflict? Oxford University Press, Oxford.

Crawley, M. J., S. L. Brown, R. S. Hails, D. D. Kohn, and M. Rees. 2001. Transgenic crops in natural habitats. Nature **409**:682-683.

Crow, J. F. 1992. An advantage of sexual reproduction in a rapidly changing environment. Journal of Heredity **83**:169-173.

Crow, J. F., and K. Kimura. 1970. An introduction to population genetics theory. Alpha Editions, Edina.

Cruden, R. W., and D. L. Lyon. 1985. Patterns of biomass allocation to male and female functions in plants with different mating systems. Oecologia **66**:299-306.

Cruden, R. W., and D. L. Lyon. 1989. Facultative xenogamy: Examination of a mixed mating system. Pages 171-207 *in* J. H. Bock and Y. B. Linhart, editors. The evolutionary ecology of plants. Westview Press, Boulder.

Dagum, C. 1980. The generation and distribution of income, the Lorenz curve and the Gini ratio. Économie Appliquée **33**:327-367.

Damgaard, C. 1996. Fixation probabilities of selfing rate modifiers in simulations with several deleterious alleles with linkage. Evolution **50**:1425-1431.

Damgaard, C. 1998. Plant competition experiments: Testing hypotheses and estimating the probability of coexistence. Ecology **79**:1760-1767.

Damgaard, C. 1999. A test of asymmetric competition in plant monocultures using the maximum likelihood function of a simple growth model. Ecological Modelling **116**:285-292.

Damgaard, C. 2000. Fixation of advantageous alleles in partially self-fertilizing populations: The effect of different selection modes. Genetics **154**:813-821.

Damgaard, C. 2002. Quantifying the invasion probability of genetically modified plants. BioSafety 7: Paper 1 (BY02001) Online Journal – URL: http://www.bioline.org.br/by.

Damgaard, C. 2003a. Evolution of advantageous alleles affecting population ecological characteristics in partially inbreeding populations. Hereditas **138**:122-128.

Damgaard, C. 2003b. Modelling plant competition along an environmental gradient. Ecological Modelling **170**:45-53.

Damgaard, C. 2004a. Dynamics in a discrete two-species competition model: coexistence and over-compensation. Journal of Theoretical Biology **227**:197-203.

Damgaard, C. 2004b. Inference from plant competition experiments: the effect of spatial covariance. Oikos **107**: 225-230.

Damgaard, C., and R. Abbott. 1995. Positive correlations between selfing rate and pollen-ovule ratio within plant populations. Evolution **49**:214-217.

Damgaard, C., D. Couvet, and V. Loeschcke. 1992. Partial selfing as an optimal mating strategy. Heredity **69**:289-295.

Damgaard, C., B. Guldbrandtsen, and F. B. Christiansen. 1994. Male gametophytic selection against a deleterious allele in a mixed mating model. Hereditas **120**:13-18.

Damgaard, C., and B. D. Jensen. 2002. Disease resistance in *Arabidopsis thaliana* increases the competitive ability and the predicted probability of long-term ecological success under disease pressure. Oikos **98**:459-466.

Damgaard, C., and V. Loeschcke. 1994a. Genotypic variation for reproductive characters, and the influence of pollen-ovule ratio on selfing rate in rape seed (*Brassica napus*). Journal of Evolutionary Biology **7**:599-607.

Damgaard, C., and V. Loeschcke. 1994b. Inbreeding depression and dominance-suppression competition after inbreeding depression in rape seed (*Brassica napus*). Theoretical and Applied Genetics **88**:321-323.

Damgaard, C., and J. Weiner. 2000. Describing inequality in plant size or fecundity. Ecology **81**:1139-1142.

Damgaard, C., J. Weiner, and H. Nagashima. 2002. Modelling individual growth and competition in plant populations: growth curves of *Chenopodium album* at two densities. Journal of Ecology **90**:666-671.

Darwin, C. R. 1859. The origin of species. John Murray, London.

Darwin, C. R. 1877. The different forms of flowers on plants of the same species. John Murray, London.

de Valpine, P., and A. Hastings. 2002. Fitting population models incorporating process noise and observation error. Ecological Monographs **72**:57-76.

de Wit, C. T. 1960. On competition. Verslagen van Landbouwkundige Onderzoekingen **66.8**:1-82.

Deutschman, D. H., S. A. Levin, C. Devine, and L. A. Buttel. 1997. Scaling from trees to forests:analysis of a complex simulation model. Science online **277**:http://www.sciencemag.org/feature/data/deutschman/index.htm.

Dickinson, C. H., and J. R. Greenhalgh. 1977. Host range and taxonomy of Peronospora on crucifers. Transactions of the British Mycological Society **69**:111-116.

Dieckmann, U., and M. Doebeli. 1999. On the origin of species by sympatric speciation. Nature **400**:354-357.

Dieckmann, U., R. Law, and J. A. J. Metz, editors. 2000. The geometry of ecological interactions: Simplifying spatial complexity. Cambridge University Press, Cambridge.

Dixon, P. M., J. Weiner, T. Mitchell-Olds, and R. Woodley. 1987. Bootstrapping the Gini coefficient of inequality. Ecology **68**:1548-1551.

Doebeli, M., and U. Dieckmann. 2000. Evolutionary branching and sympatric speciation caused by different types of ecological interactions. American Naturalist **156**:S77-S101.

Doncaster, C. P., G. E. Pound, and S. J. Cox. 2000. The ecological cost of sex. Nature **404**:281-285.

Durrett, R., and S. Levin. 1994. The importance of being discrete (and spatial). Theoretical Population Biology **46**:363-394.

Easterling, M. R., S. P. Ellner, and P. M. Dixon. 2000. Size-specific sensitivity: applying a new structured population model. Ecology **81**:694-708.

Edwards, A. W. F. 1972. Likelihood. Cambridge University Press, Cambridge.

Eguiarte, L. E., A. Búrquez, J. Rodríguez, M. Martínes-Ramos, J. Sarukhán, and D. Pinero. 1993. Direct and indirect estimates of neighborhood and effective population size in a tropical palm, *Astrocaryum mexicanum*. Evolution **47**:75-87.

Elaydi, S. N. 1999. An introduction to difference equations. Springer-Verlag, New York.

Ellison, A. M., P. M. Dixon, and J. Ngai. 1994. A null model for neighborhood models of plant competitive interactions. Oikos **71**:225-238.

Emerson, B. C., E. Paradis, and C. Thébaud. 2001. Revealing the demographic histories of species using DNA sequences. Trends in Ecology and Evolution **16**:707-716.

Ewens, W. J. 1963. The diffusion equation and a pseudo-distribution in genetics. Journal of the Royal Statistical Society, B **25**:405-412.

Ewens, W. J. 1969. Population genetics. Methuen, London.

Ewens, W. J. 1972. The sampling theory of selectively neutral alleles. Theoretical Population Biology **3**:87-112.

Feldman, M. W., and F. B. Christiansen. 1984. Population genetic theory of the cost of inbreeding. American Naturalist **123**:642-653.

Feldman, M. W., F. B. Christiansen, and U. Liberman. 1983. On some models of fertility selection. Genetics **105**:1003-1010.

Firbank, L. G., and A. R. Watkinson. 1985. On the analysis of competition within two-species mixtures of plants. Journal of Applied Ecology **22**:503-517.

Fisher, R. A. 1941. Averagde excess and average effect of a gene substitution. Annals of Eugenics **11**:53-63.

Fisher, R. A. 1958. The genetical theory of natural selection, 2 edition. Dover, New York.

Fortin, M.-J., and J. Gurevitch. 2001. Mantel tests. in S. M. Scheiner and J. Gurevitch, editors. Design and analysis of ecological experiments. Oxford University Press, Oxford.

Francis, M. G., and D. A. Pyke. 1996. Crested wheatgrass-cheatgrass seedling competition in a mixed-density design. Journal of Range Management 49:432-438.

Frank, S. A. 1997. The Price equation, Fisher's fundamental theorem, kin selection, and causal analysis. Evolution 51:1712-1729.

Freckleton, R. P., and A. R. Watkinson. 1998. Predicting the determinants of weed abundance: a model for the population dynamics of *Chenopodium album* in sugar beet. Journal of Applied Ecology 35:904-920.

Freckleton, R. P., and A. R. Watkinson. 2001. Nonmanipulative determination of plant community dynamics. Trends in Ecology and Evolution 16:301-307.

Freckleton, R. P., and A. R. Watkinson. 2002. Are weed population dynamics chaotic? Journal of Applied Ecology 39:699-707.

Gaillard, M.-J., H. J. B. Birks, M. Ihse, and S. Runborg. 1998. Pollen/landscape calibrations based on modern assemblages from surface-sediment samples and landscape mapping – a pilot study from South Sweden. Paläoklimaforschung 27:31-52.

Gale, J. S. 1990. Theoretical population genetics. Unwin Hyman, London.

Garcia-Barrios, L., D. Mayer-Foulkes, M. Franco, G. Urquido-Vasquez, and J. Franco-Perez. 2001. Development and validation of a spatial explicit individual-based mixed crop growth model. Bulletin of Mathematical Biology 63:507-526.

Gates, D. J., and M. Westcott. 1978. Zone of influence models for competition in plantations. Advances in Applied Probability 10:299-537.

Geber, M. 1989. Interplay of morphology and development on size inequality: A polygonum greenhouse study. Ecological Monographs 59:267-288.

Geritz, S. A. H. 1998. Evolutionary singular strategies and the adaptive growth and branching of the evolutionary tree. Evolutionary Ecology 12:35-57.

Geritz, S. A. H., E. van der Meijden, and J. A. J. Metz. 1999. Evolutionary dynamics of seed size and seedling competitive ability. Theoretical Population Biology 55:324-343.

Glasser, G. J. 1962. Variance formulas for the mean difference and coefficient of concentration. Journal of American Statistical Association 57:648-654.

Goldberg, D. E., and A. M. Barton. 1992. Patterns and consequences of interspecific competition in natural communities: A review of field experiments with plants. American Naturalist **139**:771-801.

Golding, B., editor. 1994. Non-neutral evolution. Chapman & Hall, New York.

Gotelli, N., and D. J. McCabe. 2002. Species co-occurrence: a meta-analysis of J. M. Diamonds's assembly rules model. Ecology **83**:2091-2096.

Gouyon, P.-H., and D. Couvet. 1987. A conflict between two sexes, females and hermaphrodites. *In* S. C. Sterns, editor. The evolution of sex and its consequences. Birkhäuser, Basel.

Gregorius, H.-R. 1982. Selection in plant populations of effectively infinite size. II. Protectedness of a biallelic polymorphism. Journal of Theoretical Biology **96**:689-705.

Gregorius, H.-R., M. Ziehe, and M. D. Ross. 1987. Selection caused by self-fertilisation. I. Four measures of self-fertilisation and their effects on fitness. Theoretical Population Biology **31**:91-115.

Greiner La Peyre, M. K., E. Hahn, I. A. Mendelssohn, and J. B. Grace. 2001. The importance of competition in regulating plant species abundance along a salinity gradient. Ecology **82**:62-69.

Grime, P. 2001. Plant Strategies, Vegetation Processes, and Ecosystem Properties, 2 edition. Wiley.

Gurevitch, J., L. L. Morrow, A. Wallace, and J. S. Walsh. 1992. A meta-analysis of competition in field experiments. American Naturalist **140**:539-572.

Gurtin, M. E., and R. C. MacCamy. 1979. Some simple models for nonlinear age-dependent population dynamics. Mathematical Biosciences **43**:199-211.

Haldane, J. B. S. 1924. A mathematical theory of natural and artificial selection. Part II. Proceedings of the Cambridge Philological Society – Biological Science **1**:158-163.

Haldane, J. B. S. 1927. A mathematical theory of natural and artificial selection. V. Selection and mutation. Proceedings of the Cambridge Philological Society **23**:838-844.

Haldane, J. B. S. 1932. The causes of evolution. Harper, New York.

Hamilton, W. D. 1993. Haploid dynamic polymorphism in a host with mathching parasites: Effects of mutation/subdivision, linkage, and patterns of selection. Journal of Heredity **84**:328-338.

Hanski, I. 1998. Metapopulation dynamics. Nature **396**:41-49.

Hara, T., and T. Wyszomirski. 1994. Competitive asymmetry reduces spatial effects on size-structure dynamics in plant populations. Annals of Botany **73**:173-190.

Harper, J. L. 1977. Population biology of plants. Academic Press, London.

Hartl, D. L., and A. G. Clark. 1989. Principles of population genetics. Sinauer, Sunderland.

Harushima, Y., M. Nakagahra, M. Yano, T. Sasaki, and N. Kurata. 2001. A genome-wide survey of reproductive barriers in an intraspecific hybrid. Genetics **159**:883-892.

Haskell, E. 1947. A natural classification of societies. New York Academy of Science, Trans. Series 2 **9**:186-196.

Hassell, M. P., and H. N. Comins. 1976. Discrete time models for two-species competition. Theoretical Population Biology **9**:202-221.

Heino, M., J. A. J. Metz, and V. Kaitala. 1998. The enigma of frequency-dependent selection. TREE **13**:367-370.

Hiebeler, D. 1997. Stochastic spatial models: From simulations to mean field and local structure approximation. Journal of Theoretical Biology **187**:307-319.

Hofbauer, F., J. Hofbauer, P. Raith, and T. Steinberger. 2004. Intermingled basins in a two species system. Journal of Mathematical Biology **49**:293-309.

Hofbauer, J., and K. Sigmund. 1998. Evolutionary games and population dynamics. Cambridge University Press, Cambridge.

Holsinger, K. E. 1991. Mass-action models of plant mating systems: the evolutionary stability of mixed mating systems. American Naturalist **138**:606-622.

Holsinger, K. E. 1996. Pollination biology and the evolution of mating systems in flowering plants. Pages 107-149 *in* Evolutionary Biology. Plenum Press, New York.

Holsinger, K. E. 2000. Reproductive systems and evolution in vascular plants. Proceedings of the National Academy of Science **97**:7037-7042.

Holt, R. D. 1977. Predation, apparent competition and the structure of prey communities. Theoretical Population Biology **12**:197-229.

Holub, E. B., and J. L. Beynon. 1997. Symbiology of mouse-ear cress (*Arabidopsis thaliana*) and Oomycetes. Advances in Botanical Research **24**:227-273.

Holub, E. B., J. L. Beynon, and I. R. Crute. 1994. Phenotypic and genotypic characterization of interactions between isolates of *Peronospora parasitica* and accessions of *Arabidopsis thaliana*. Molecular Plant-Microbe Interactions **7**:223-239.

Hooper, D. U. 1998. The role of complementarity and competition in ecosystem resonses to variation in plant diversity. Ecology **79**:704-719.

Hoppensteadt, F. C. 1982. Mathematical methods of population biology. Cambridge University Press, Cambridge.

Hormaza, J. I., and M. Herrero. 1992. Pollen selection. Theoretical and Applied Genetics **83**:663-672.

Hudson, P., and J. Greenman. 1998. Competition mediated by parasites: Biological and theoretical progress. Trends in Ecology and Evolution 13:387-390.

Huisman, J., and F. J. Weissing. 2001. Biological conditions for ocsilations and chaos generated by multispecies competition. Ecology 82:2682-2695.

Inouye, B. 2001. Response surface experimental designs for investigating interspecific competition. Ecology 82:2696-2706.

Inouye, B., and W. M. Schaffer. 1981. On the ecological meaning of ratio (de Wit) diagrams in plant ecology. Ecology 62:1679-1681.

Jarry, M., M. Khaladi, M. Hossaert-McKey, and D. McKey. 1995. Modelling the population dynamics of annual plants with seed bank and density dependent effects. Acta Biotheoretica 43:53-65.

Joshi, A. 1997. Laboratory studies of density-dependent selection: Adaptations to crowding in *Drosophila melanogaster*. Current Science 72:555-562.

Kareiva, P., I. M. Parker, and M. Pascual. 1996. Can we use experiments and models in predicting the invasiveness of genetically engineered organisms? Ecology 77:1670-1675.

Kawabe, A., K. Yamane, and N. T. Miyashita. 2000. DNA polymorphism at the cytosolic phosphoglusoce isomerase (PgiC) locus of the wild plant Arabidopsis thaliana. Genetics 156:1339-1347.

Keddy, P. A. 1990. Competitive hierachies and centrifugal organisation in plant communities. *In* J. B. Grace and D. Tilman, editors. Perspectives on plant competition. Academic Press, London.

Kimura, M. 1962. On the probability of fixation of mutant genes in a population. Genetics 47:713-719.

Kimura, M. 1983. The neutral theory of molecular evolution. Cambridge University Press, Cambridge.

Kjellsson, G., and V. Simonsen. 1994. Methods for risk assessment of transgenic plants. I. Competition, establishment and ecosystems effects. Birkhäuser, Basel.

Kjellsson, G., V. Simonsen, and K. Ammann, editors. 1997. Methods for risk assessment of transgenic plants. II. Pollination, gene-transfer and population impacts. Birkhäuser, Basel.

Knox, R. G., R. K. Peet, and N. L. Christiensen. 1989. Population dynamics in loblolly pine stands: changes in skewness and size inequality. Ecology 70:1153-1166.

Kot, M., M. A. Lewis, and P. van den Driessche. 1996. Dispersal data and the spread of invading organisms. Ecology 77:2027-2042.

Kotz, S., N. L. Johnson, and C. B. Read. 1983. Encyclopedia of statistical science. John Wiley & Sons, New York.

Krivan, V. 1996. Optimal foraging and prdator-prey dynamics. Theoretical Population Biology 49:265-290.

Krivan, V., and A. Sidker. 1999. Optimal foraging and predator-prey dynamics. Theoretical Population Biology **55**:111-126.

Kuhner, M. K., J. Yamato, and J. Felsenstein. 1995. Estimating effective population size and mutation rate from sequence data using Metropolis-Hastings sampling. Genetics **140**:1421-1430.

Lande, R. 1993. Risks of population extinction from demographic and environmental stochasticity and random catastrophes. American Naturalist **142**:911-927.

Lande, R., and S. J. Arnold. 1983. The measurement of selection on correlated characters. Evolution **37**:1210-1226.

Lande, R., and D. W. Schemske. 1985. The evolution of self-fertilisation and inbreeding depression in plants. I. Genetic models. Evolution **39**:24-40.

Law, R., P. Marrow, and U. Dieckmann. 1997. On evolution under asymmetric competition. Evolutionary Ecology **11**:485-501.

Law, R., and R. D. Morton. 1996. Permanence and the assembly of ecological communities. Ecology **77**:762-775.

Law, R., and A. R. Watkinson. 1987. Response-surface analysis of two species competition: An experiment on *Phleum arenarium* and *Vulpia fasciculata*. Journal of Ecology **75**:871-886.

Levin, S. A., B. Grenfell, A. Hastings, and A. S. Perelson. 1997. Mathematical and computational challenges in population biology and ecosystems science. Science **275**:334-343.

Levin, S. A., and H. C. Muller-Landau. 2000. The evolution of dispersal and seed size in plant communities. Evolutionary Ecology Research **2**:409-435.

Levin, S. A., and S. W. Pacala. 1997. Theories of simplification and scaling of spatially distributed processes. Pages 271-295 *in* D. Tilman and P. Kareiva, editors. Spatial Ecology. Princeton University Press, Princeton.

Lewis, D. 1941. Male-sterility in natural populations of hermaphrodite plants. New Phytology **40**:56-63.

Lewontin, R. C. 1970. The units of selection. Review of Ecology and Systematics **1**:1-18.

Lloyd, D. G., and K. S. Bawa. 1984. Modification of the gender of seed plants in varying conditions. Evolutionary Biology **17**:255-338.

Lomnicki, A. 2001. Carrying capacity, competition and maintenance of sexuality. Evolutionary Ecology Research **3**:603-610.

MacArthur, R. H., and R. Levins. 1967. The limiting similarity, convergence and divergence of coexisting species. American Naturalist **101**:377-385.

Manly, B. F. J. 1985. The statistics of natural selection on animal populations. Chapman and Hall, London.

Marshall, D. R., and S. K. Jain. 1969. Interference in pure and mixed populations of *Avena fatua* and *A. barbata*. Journal of Ecology **57**:251-270.

May, R. M., and R. M. Anderson. 1983. Epidemiology and genetics in the coevolution of parasites and hosts. Proceedings of the Royal Society of London, **B 219**:281-313.

May, R. M., and G. F. Oster. 1976. Bifurcations and dynamic complexity in simple ecological models. American Naturalist **110**:573-599.

Maynard Smith, J. 1978. The evolution of sex. Cambridge University Press, Cambridge.

Maynard Smith, J., and G. R. Price. 1973. The logic of animal conflict. Nature **246**:15-18.

Mayr, E. 1942. Systematics and the origin of species. Columbia University Press, New York.

Mazer, S. J. 1987. Parental effects on seed development and seed yield in *Raphanus raphanistrum*: Implications for natural and sexual selection. Evolution **41**:355-371.

McDonald, J. H., and M. Kreitman. 1991. Adaptive protein evolution at the *Adh* locus in *Drosophila*. Nature **351**:652-654.

Mead, R. 1967. A mathematical model for the estimation of inter-plant competition. Biometrics **23**:189-205.

Mead, R. 1970. Plant density and crop yield. Applied statistics **19**:64-81.

Mitchell-Olds, T., and J. Bergelson. 1990. Statistical genetics of an annual plant, *Impatiens capensis*. II. Natural selection. Genetics **124**:417-421.

Mitchell-Olds, T., and R. G. Shaw. 1987. Regression analysis of natural selection: statistical inference and biological interpretation. Evolution **41**:1149-1161.

Miyashita, N. T., A. Kawabe, and H. Innan. 1999. DNA variation in the wild plant *Arabidopsis thaliana* revealed by amplified fragment length polymorphism analysis. Genetics **152**:1723-1731.

Mulcahy, D. L., M. Sari-Gorla, and G. B. Mulcahy. 1996. Pollen selection – past, present and future. Sexual Plant Reproduction **9**:353-356.

Nagashima, H., I. Terashima, and S. Katoh. 1995. Effects of plant density on frequency distributions of plant height in *Chenopodium album* strands: Analysis based on continous monitoring of height-growth of individual plants. Annals of Botany **75**:173-180.

Nagylaki, T. 1998. Fixation indices in subdivided populations. Genetics **148**:1325-1332.

Nassar, J. M., J. L. Hamrick, and T. H. Flemming. 2001. Genetic variation and population structure of the mixed-mating cactus, *Melocactus curvispinus* (Cactaceae). Heredity **87**:69-79.

Nei, M. 1987. Molecular evolutionary genetics. Columbia University Press, New York.

Neubert, M. G., and H. Caswell. 2000. Demography and dispersal: Calculation of sensitivity analysis of invasion speed for structured populations. Ecology **81**:1613-1628.

Nielsen, R. 2001. Statistical tests of selective neutrality in the age of genomics. Heredity **86**:641-647.

Nordborg, M., and P. Donnelly. 1997. The coalescent process with selfing. Genetics **146**:1185-1195.

Nordborg, M., and H. Innan. 2002. Molecular population genetics. Current Opinion in Plant Biology **5**:69-73.

Nunney, L. 2002. The effective size of annual plant populations: the interaction of a seed bank with fluctuating population size in maintaining genetic variation. American Naturalist **160**:195-204.

Odgaard, B., P. Eigaard, A. B. Nielsen, and J. R. Rømer. 2001. AGRAR 2000. Det agrare landskab fra Kristi fødsel til det 21. århundrede: Kvantitative estimater, regionalitet og årsager til forandringer. Pages 51-75 *in* Forskningsrådene, editor. Det Agrare Landskab 1998-2002, Midtvejsrapport. Forskningsstyrelsen, København.

Ohta, T., and C. C. Cockerham. 1974. Detrimental genes with partial selfing and effects on a neutral locus. Genetical Research, Cambridge **23**:191-200.

Otte, D., and J. A. Endler, editors. 1989. Speciation and its consequences. Sinauer Associates, Sunderland, Mass.

Otto, S. P., and M. C. Whitlock. 1997. The probability of fixation in populations of changing size. Genetics **146**:723-733.

Overath, R. D., and M. Asmussen. 1998. Genetic diversity at a single locus under viability selection and facultative apomixis: Equilibrium structure and deviations from Hardy-Weinberg frequencies. Genetics **148**:2029-2039.

Pacala, S., C. Canham, J. Saponara, J. Silander, R. Kobe, and E. Ribbens. 1996. Forest models defined by field measurements: Estimation, error analysis and dynamics. Ecological Monographs **66**:1-44.

Pacala, S., and S. A. Levin. 1997. Biological generated spatial pattern and the coexistence of competing species. *In* D. Tilman and P. Kareiva, editors. Spatial ecology. The role of space in population dynamics and interspecific interactions. Princeton University Press, Princeton.

Pacala, S. W., and J. A. Silander. 1985. Neighboorhood models of plant population dynamics. I. Single species models of annuals. American Naturalist **125**:385-411.

Pacala, S. W., and J. A. Silander. 1987. Neighborhood interference among velvet leaf, *Abutilon theophrasti*, and pigweed, *Amaranthus retroflexus*. Oikos **48**:217-224.

Pacala, S. W., and J. A. Silander. 1990. Field tests of neighborhood population dynamic models of two annual species. Ecological Monographs **60**:113-134.

Pannel, J. R., and B. Charlesworth. 2000. Effects of metapopulation processes on measures of genetic diversity. Philosophical Transactions of the Royal Society, London Series **B** 355:1851-1864.

Partridge, L., and L. D. Hurst. 1998. Sex and conflict. Science **281**:2003-2008.

Penrose, L. S. 1949. The meaning of "fitness" in human populations. Annals of Eugenics **14**:301-304.

Pigliucci, M., K. Cammell, and J. Schmitt. 1999. Evolution of phenotypic plasticity a comparative approach in the phylogenetic neighbourhood of *Arabidopbsis thaliana*. Journal of Evolutionary Biology **12**:779-791.

Pollak, E., and M. Sabran. 1992. On the theory of partially inbreeding finite population. III. Fixation probabilities under partial selfing when heterozygotes are intermediate in viability. Genetics **131**:979-985.

Preston, K. A. 1998. Architectural constrients on flower number in a photoperiodic annual. Oikos **81**:279-288.

Prout, T. 1965. The estimation of fitnesses from genotypic frequencies. Evolution **19**:546-551.

Queller, D. C. 1987. Sexual selection in flowering plants. *In* J. W. Bradbury and M. B. Andersson, editors. Sexual selection: Testing the alternatives. John Wiley, New York.

Rees, M., R. Condit, M. Crawley, S. Pacala, and D. Tilman. 2001. Long-term studies of vegetation dynamics. Science **293**:650-655.

Rees, M., P. J. Grubb, and D. Kelly. 1996. Quantifying the impact of competition and spatial heterogeneity on the structure and dynamics of a four-species guild of winter annuals. American Naturalist **147**:1-32.

Rees, M., and R. L. Hill. 2001. Large-scale disturbances, biological control and the dynamics of gorse populations. Journal of Applied Ecology **38**:364-377.

Rees, M., D. Kelly, and O. N. Bjørnstad. 2002. Snow tussocks, chaos, and the evolution of mast seeding. American Naturalist **160**:44-59.

Rees, M., and M. J. Long. 1993. The analysis and interpretation of seedling recruitment curves. American Naturalist **141**:233-262.

Rees, M., and Q. Paynter. 1997. Biological control of Scotch broom: modelling the determinats of abundance and the potential impact of introduced insect herbivores. Journal of Applied Ecology **34**:1203-1221.

Richards, F. J. 1959. A flexible growth function for empirical use. Journal of Experimental Botany **10**:290-300.

Rocheleau, G., and S. Lessard. 2000. Stability analysis of the partial selfing selection model. Journal of Mathematical Biology **40**:541-574.

Ross, M. 1990. Sexual asymmetry in hermaphroditic plants. Trends in Ecology and Evolution 5:43-47.

Roxburgh, S. H., and J. B. Wilson. 2000a. Stability and coexistence in a lawn community: experimental assessment of the stability of the actual community. Oikos **88**:309-423.

Roxburgh, S. H., and J. B. Wilson. 2000b. Stability and coexistence in a lawn community: mathematical prediction of stability using a community matrix with parameters derived from competition experiments. Oikos **88**:395-408.

Schemske, D. W., and R. Lande. 1985. The evolution of self-fertization and inbredding depression in plants. II. Empirical observations. Evolution **39**:41-52.

Schlichting, C. D., and B. Devlin. 1989. Male and female reproductive success in the hemaphroditic plant *Phlox drummondii*. American Naturalist **133**:212-227.

Schluter, D. 2000. The ecology of adaptive radiation. Oxford University Press, Oxford.

Schluter, D., and D. Nychka. 1994. Exploring fitness surfaces. American Naturalist **143**:597-616.

Schmid, B., W. Polasek, J. Weiner, A. Krause, and P. Stoll. 1994. Modelling of discontinuos relationships in biology with censored regression. American Naturalist **143**:494-507.

Schoen, D. J., and D. G. Lloyd. 1992. Self- and cross-fertilization in plants. III. Methods for studying modes and functional aspects of self-fertilization. International Journal of Plant Science **153**:381-393.

Schwinning, S., and G. A. Fox. 1995. Population dynamic consequences of competitive symmetry in annual plants. Oikos **72**:422-432.

Schwinning, S., and J. Weiner. 1998. Mechanisms determining the degree of size asymmetry in competition among plants. Oecologia **113**:447-455.

Seber, G. A. F., and C. J. Wild. 1989. Nonlinear regression. John Wiley & Sons, New York.

Selgrade, J. F., and M. Ziehe. 1987. Convergences to equilibrium in a genetic model with differential viability between the sexes. Journal of Mathematical Biology **25**:477-490.

Sen, A. 1973. On economic inequality. Clarendon, Oxford.

Shigesada, N., and K. Kawasaki. 1997. Biological invasions: Theory and practice. Oxford University Press, Oxford.

Shinozaki, K., and T. Kira. 1956. Intraspecific competition among higher plants. VII. Logistic theory of the C-D effect. J. Inst. Polytech, Osaka City University **7**:35-72.

Shumway, D. L., and R. T. Koide. 1995. Size and reproductive inequality in mycorrhizal and nonmycorrhizal populations of *Abutilon theophrasti*. Journal of Ecology **83**:613-620.

Silvertown, J., and D. Charlesworth. 2001. Introduction to plant population biology. Blackwell Science, Oxford.

Silvertown, J., M. E. Dodd, D. J. G. Gowing, and J. O. Mountford. 1999. Hydrologically defined niches reveal a basis for species richness in plant communities. Nature **400**:61-63.

Skovgaard, I. 1986. A statistical model for competition experiments. Scandinavian Journal of Statistics **13**:29-38.

Slatkin, M. 1979. The evolutionary response to frequency- and density-dependent interactions. American Naturalist **114**:384-398.

Sletvold, N., and G. Hestmark. 1999. A comparative test of the predictive power of neighbourhood models in natural populations of *Lasillia pustulata*. Canadian Journal of Botany **77**:1655-1661.

Smith, C. C., and S. D. Fretwell. 1974. The optimal balance between size and number of offspring. American Naturalist **108**:499-506.

Smithson, A., and M. R. Magnar. 1997. Density-dependent and frequency-dependent selection by bumblebees *Bombus terrestris* (L.) (Hymenoptera: Apidae). Biological Journal of the Linnean Society **60**:401-417.

Soares, P., and M. Tomé. 1999. Distance-dependent competition measures for eucalyptus plantations in Portugal. Annals of Forest Science **56**:307-319.

Stebbins, G. L. 1950. Variation and evolution in plants. Columbia University Press, New York.

Stephenson, A. G., and R. I. Bertin. 1983. Male competition, female choice, and sexual selection in plants. Pages 109-149 *in* L. Real, editor. Pollination biology. Academic Press, New York.

Stockmarr, A. 2002. The distribution of particles in the plane dispersed by a simple 3-dimensional difussion process. Journal of Mathematical Biology.

Stokes, K. E., J. M. Bullock, and A. R. Watkinson. 2004. Population dynamics across a parapatric range boundary: *Ulex gallii* and *Ulex minor*. Journal of Ecology **92**:142-155.

Stoll, P., and D. Prati. 2001. Intraspecific aggregation alters competitive interactions in experimental plant populations. Ecology **82**:319-327.

Stoll, P., and J. Weiner. 2000. A neighborhood view of interactions among individual plants. Pages 11-27 *in* U. Dieckmann, R. Law, and J. A. J. Metz, editors. The geometry of ecological interactions. Cambridge University Press, Cambridge.

Stratton, D. A. 1995. Spatial scale of variation in fitness of *Erigeron annuus*. American Naturalist **146**:608-624.

Stuart, A., and J. K. Ord. 1991. Kendall's advanced theory of statistics, 5 edition. Edward Arnold, London.

Symstad, A. J. 2000. A test of the effect of functional group richness and composition on grassland invasibility. Ecology 81:99-109.

Tilman, D. 1988. Dynamics and structure of plant communities. Princeton University Press, Princeton.

Tilman, D. 1994. Competition and biodiversity in spatially structured habitats. Ecology 75:2-16.

Tilman, D. 1997. Community invasibility, recruitment limitation, and grassland biodiversity. Ecology 78:81-92.

Tilman, D., and P. Kareiva, editors. 1997. Spatial ecology. The role of space in population dynamics and interspecific interactions. Princeton University Press, Princeton.

Uyenoyama, M., K. E. Holsinger, and D. M. Waller. 1993. Ecological and genetic factors directing the evolution of self-fertilisation. Pages 327-379 in Oxford surveys in evolutionary biology.

Uyenoyama, M. K., and D. M. Waller. 1991a. Coevolution of self-fertilisation and inbreeding depression. I. Mutation-selection balance at one and two loci. Theoretical Population Biology 40:14-46.

Uyenoyama, M. K., and D. M. Waller. 1991b. Coevolution of self-fertilisation and inbreeding depression. II. Symmetric overdominance in viability. Theoretical Population Biology 40:47-77.

Uyenoyama, M. K., and D. M. Waller. 1991c. Coevolution of self-fertilisation and inbreeding depression. III. Homozygous lethal mutations at multiple loci. Theoretical Population Biology 40:173-210.

Valdes, A. M., M. Slatkin, and N. B. Freimer. 1993. Allele frequencies at microsatellite loci: The stepwise mutation model revisited. Genetics 133:737-749.

Vandermeer, J. 1984. Plant competition and the yield-density relationship. Journal of Theoretical Biology 109:393-399.

Vandermeer, J. H. 1989. The ecology of intercropping. Cambridge University Press, Cambridge.

Vaupel, W. V., K. G. Manton, and E. Stallard. 1979. The impact of heterogeneity in individual frailty on the dynamics of mortality. Demography 16:439-454.

Verhulst, J. H. 1838. Notice sur la loi que population suit dans son accroissement. Correspondance mathématique et physique 10:113-121.

Walsh, N. E., and D. Charlesworth. 1992. Evolutionary interpretations of differences in pollen tube growth rates. Quarterly Review of Biology 67:19-37.

Wang, J. 1997. Effective size and F-statistics of subdivided populations. I. Monoecious species with partial selfing. Genetics 146:1453-1463.

Watkinson, A. R. 1980. Density-dependence in single species populations of plants. Journal of Theoretical Biology **83**:345-357.

Watkinson, A. R., R. P. Freckleton, and L. Forrester. 2000. Population dynamics of *Vulpia ciliata*: regional, patch and local dynamics. Journal of Ecology **88**:1012-1029.

Watterson, G. A. 1977. Heterosis or neutrality? Genetics **85**:789-814.

Weiher, E., G. D. P. Clarke, and P. A. Keddy. 1998. Community assembly rules, morphological dispersion, and the coexistence of plant species. Oikos **81**:309-322.

Weiner, J. 1984. Neighbourhood interference amongst *Pinus rigida* individuals. Journal of Ecology **72**:183-195.

Weiner, J. 1985. Size hierarchies in experimental populations of annual plants. Ecology **66**:743-752.

Weiner, J. 1986. How competition for light and nutrients affects size variability in *Ipomoea tricolor* populations. Ecology **67**:1425-1427.

Weiner, J. 1990. Asymmetric competition in plant populations. Trends in Ecology and Evolution **5**:360-364.

Weiner, J., and O. T. Solbrig. 1984. The meaning and measurement of size hierarchies in plant populations. Oecologia **61**:334-336.

Weiner, J., P. Stoll, H. Muller-Landau, and A. Jasentuliyana. 2001. The effects of density, spatial pattern and competitive symmetry on size variation in simulated plant populations. American Naturalist **158**:438-450.

Weir, B. S. 1996. Genetic data analysis, 2 edition. Sinauer, Sunderland.

Westoby, M. 1998. A leaf-height-seed (LHS) plant ecological strategy scheme. Plant and Soil **199**:213-227.

Westoby, M., E. Jurado, and M. Leishman. 1992. Comparative evolutionary ecology of seed size. Trends in Ecology and Evolution **7**:368-372.

Whitlock, M. C., and N. H. Barton. 1997. The effective size of a subdivided population. Genetics **146**:427-441.

Wilson, J. B., M. J. Crawley, M. E. Dodd, and J. Silvertown. 1996. Evidensce for constraint on species coexistence in vegetation of the Park Grass experiment. Vegetatio **124**:183-190.

Winn, A. A., and T. E. Miller. 1995. Effect of density on magnitude of directional selection on seed mass and emergence time in *Plantago wrightiana* Dcne. (Plantaginaceae). Oecologia **103**:365-370.

Wright, S. 1969. Evolution and the genetics of populations. Vol. 2. The theory of gene frequencies. The University of Chicago Press, Chicago.

Wyszomirski, T. 1983. A simulation model of the growth of competing individuals of a plant population. Ekologia Polska **31**:73-92.

Wyszomirski, T., I. Wyszomirska, and I. Jarzyna. 1999. Simple mechanisms of size distributions dynamics in crowded and uncrowded virtual monocultures. Ecological Modelling 115:253-273.

Yampolsky, E., and H. Yampolsky. 1922. Distribution of sex forms in the phanerogamic flora. Bibliotheca Genetica 3:1-62.

Yan, G. 1996. Parasite-mediated competition: A model of directly transmitted macroparasites. American Naturalist 148:1089-1112.

Yoda, K., T. Kira, H. Ogawa, and K. Hozumi. 1963. Self-thinning in overcrowded pure stands under cultivated and natural conditions. Journal of Biology, Osaka City University 14:107-129.

Index